水产动物氨基酸营养研究

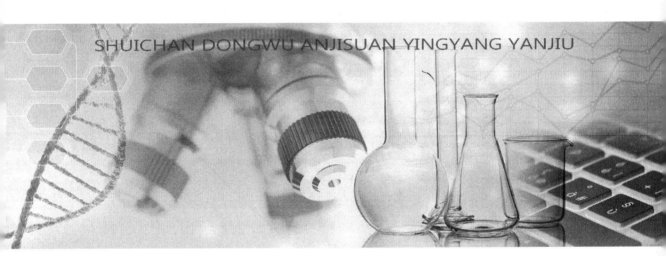

SHUICHAN DONGWU ANJISUAN YINGYANG YANJIU

王连生 徐奇友 主编

中国农业出版社
农村读物出版社
北京

本书编审人员

主　编　王连生（中国水产科学研究院黑龙江水产研究所）
　　　　　徐奇友（湖州师范学院）
副主编　范　泽（中国水产科学研究院黑龙江水产研究所）
　　　　　吴　迪（中国水产科学研究院黑龙江水产研究所）
　　　　　刘红柏（中国水产科学研究院黑龙江水产研究所）
参　编　李晋南（中国水产科学研究院黑龙江水产研究所）
　　　　　张圆圆（东北林业大学）
　　　　　孟庆维（东北农业大学）
　　　　　杨　喆（湖州师范学院）
　　　　　王常安（中国水产科学研究院黑龙江水产研究所）
　　　　　程镇燕（天津农学院）
　　　　　王庆奎（天津农学院）
　　　　　韩世成（中国水产科学研究院黑龙江水产研究所）
　　　　　刘　洋（中国水产科学研究院黑龙江水产研究所）
　　　　　陆绍霞（中国水产科学研究院黑龙江水产研究所）
　　　　　曹顶臣（中国水产科学研究院黑龙江水产研究所）
　　　　　缪凌鸿（中国水产科学研究院淡水渔业研究中心）
主　审　解绶启（中国科学院水生生物研究所）
　　　　　戈贤平（中国水产科学研究院淡水渔业研究中心）

前言
FOREWORD

随着我国水产养殖业的迅速发展，水产饲料行业也取得了长足进步，但饲料原料短缺，特别是蛋白质原料短缺，已经成为水产饲料行业发展的瓶颈问题。鱼粉作为水产动物饲料最重要的蛋白质原料之一，我国年产量约为40万吨，每年进口量达140万吨以上；豆粕作为我国第二大饲料蛋白源，供应也十分紧张，我国年产大豆约为1 500万吨，但进口量却达1亿吨左右。因此，开展饲料蛋白原料开发、提高水产动物蛋白质沉积率以及使用杂粮替代鱼粉、豆粕研究已经成为水产动物营养与饲料科学研究的热点和难点。

蛋白质是生命的物质基础之一，也是水产动物必需的重要营养物质。蛋白质经消化，转变为氨基酸和小肽，进一步被吸收后发挥其营养生理功能。因此，水产动物对蛋白质的营养需要，实质是对氨基酸的营养需要。本书介绍了氨基酸的概况及需要量的研究方法，重点介绍了水产动物必需氨基酸赖氨酸、蛋氨酸、苏氨酸、精氨酸、亮氨酸、异亮氨酸、缬氨酸、色氨酸、苯丙氨酸、组氨酸的代谢途径、营养生理功能、需要量及影响因素，并就必需氨基酸对水产动物生长、蛋白质沉积、免疫功能、肠道健康以及其他营养素代谢的影响进行了介绍。此外，还对非必需功能性氨基酸牛磺酸、谷氨酰胺及多肽的代谢途径、生理功能及作用机制进行了阐述，并就氨基酸的应用与检测进行了介绍。为提高我国水产动物蛋白质氨基酸营养与饲料理论水平，促进行业发展，我们在总结国内外近年来最新研究成果的基础上，编写了此书。本书主要适用于高等院校、科研院所科技工作者，饲料企业营养配方师、技术服务人员及养殖企业从业者阅读。

本书第一、二、十六章由徐奇友编写，第三章由范泽、王常安和王连生编写，第四章由王连生、范泽和刘红柏编写，第五章由范泽、张圆圆、陆绍霞和王连生编写，第六章由范泽、王连生和吴迪编写，第七章由张圆圆、曹顶臣和王连生编写，第八章由李晋南、王连生和范泽编写，第九章由张圆圆、王常安和王连生编写，第十章由王连生、吴迪和刘红柏编写，第十一章由吴迪、范泽和程镇燕编写，第十二章由孟庆维、杨喆、吴迪和王连生编写，第十三章由范泽、王庆奎和王连生编写，第十四章由吴迪、刘洋和韩世成编写，第十五章由吴迪、缪凌鸿和王连生编写。

感谢国家重点研发计划"蓝色粮仓科技创新"重点专项（2019YFD0900200）、国家自然科学基金（31972800、31802305）、财政部和农业农村部国家现代农业产业技术体系（CARS-45、CARS-46）、中国水产科学研究院基本科研业务费（2022XT0402）等项目的资助。

由于编者时间和水平有限，书中难免存在疏漏和错误之处，敬请读者批评指正，以臻完善。

编　者

2022 年 6 月

目 录
CONTENTS

前言

第一篇 氨基酸的概况及需要量研究方法

第一章 氨基酸的概况 ········· 3
- 第一节 氨基酸的结构 ········· 3
- 第二节 氨基酸的营养功能 ········· 4
- 第三节 氨基酸的代谢 ········· 5
  - 一、氨基酸的吸收 ········· 5
  - 二、氨基酸降解 ········· 6
  - 三、氨基酸的合成 ········· 7
- 第四节 氨基酸的营养分类 ········· 8
- 参考文献 ········· 8

第二章 水产动物氨基酸需要量研究方法 ········· 10
- 一、生长试验法 ········· 10
- 二、全鱼氨基酸组成 ········· 10
- 三、氨基酸需要量的析因方法 ········· 11
- 四、血液和肌肉中的氨基酸浓度 ········· 12
- 五、氨基酸氧化研究 ········· 12
- 参考文献 ········· 13

第二篇 水产动物必需氨基酸营养

第三章 赖氨酸 ········· 17
- 第一节 赖氨酸的理化性质 ········· 17
- 第二节 水产动物赖氨酸需要量 ········· 17

第三节　水产动物赖氨酸需要量及利用的影响因素 …………………………………… 21
　　　　一、品种、食性、生长阶段 ………………………………………………………… 21
　　　　二、饲料因素 ………………………………………………………………………… 21
　　　　三、赖氨酸的添加形式 ……………………………………………………………… 22
　　　　四、投喂策略 ………………………………………………………………………… 24
　　　　五、水温 ……………………………………………………………………………… 24
　　第四节　赖氨酸与水产动物健康 ………………………………………………………… 24
　　　　一、赖氨酸对水产动物消化吸收能力的影响 ……………………………………… 24
　　　　二、赖氨酸对水产动物免疫功能的影响 …………………………………………… 25
　　　　三、赖氨酸对水产动物氮代谢的影响 ……………………………………………… 26
　　　　四、赖氨酸对脂肪代谢及糖代谢的影响 …………………………………………… 26
　　第五节　小结与展望 ……………………………………………………………………… 27
　　参考文献 …………………………………………………………………………………… 27

第四章　蛋氨酸 ………………………………………………………………………………… 33

　　第一节　蛋氨酸的理化性质 ……………………………………………………………… 33
　　第二节　蛋氨酸的代谢途径 ……………………………………………………………… 33
　　第三节　水产动物对蛋氨酸的需要量 …………………………………………………… 34
　　第四节　蛋氨酸与其他营养素的相互作用 ……………………………………………… 36
　　　　一、蛋氨酸与牛磺酸之间的相互作用 ……………………………………………… 36
　　　　二、蛋氨酸与赖氨酸之间的相互作用 ……………………………………………… 36
　　　　三、蛋氨酸与其他营养素的相互作用 ……………………………………………… 37
　　第五节　不同类型蛋氨酸的生物效价 …………………………………………………… 37
　　　　一、包膜蛋氨酸的生物效价 ………………………………………………………… 37
　　　　二、蛋氨酸羟基类似物及其钙盐的生物效价 ……………………………………… 38
　　　　三、其他蛋氨酸的生物效价 ………………………………………………………… 39
　　第六节　蛋氨酸对水产动物蛋白质代谢及健康的影响 ………………………………… 39
　　　　一、蛋氨酸对机体蛋白质代谢的影响 ……………………………………………… 39
　　　　二、蛋氨酸对肠道健康的影响 ……………………………………………………… 40
　　　　三、蛋氨酸对机体免疫、抗氧化功能的影响 ……………………………………… 40
　　第七节　小结与展望 ……………………………………………………………………… 41
　　参考文献 …………………………………………………………………………………… 41

第五章　苏氨酸 ………………………………………………………………………………… 47

　　第一节　苏氨酸的理化性质 ……………………………………………………………… 47
　　第二节　水产动物苏氨酸需要量 ………………………………………………………… 47
　　第三节　苏氨酸对水产动物肉质的影响 ………………………………………………… 50
　　第四节　苏氨酸对水产动物蛋白质代谢的影响 ………………………………………… 50

第五节	苏氨酸对水产动物消化吸收能力的影响	50
第六节	苏氨酸对水产动物抗氧化能力的影响	51
第七节	苏氨酸对水产动物免疫功能的影响	52
第八节	小结与展望	53
参考文献		53

第六章　精氨酸 ································· 57

第一节　精氨酸的理化特性及在体内的代谢途径 ································· 57
第二节　精氨酸需要量及其对生长性能的影响 ································· 58
第三节　精氨酸对鱼类体成分的影响 ································· 59
　　一、精氨酸对鱼体粗蛋白、粗脂肪含量的影响 ································· 59
　　二、精氨酸对鱼体氨基酸组成的影响 ································· 60
第四节　精氨酸对鱼类免疫功能的影响 ································· 60
　　一、精氨酸对免疫系统的促进作用 ································· 60
　　二、精氨酸的抗氧化作用 ································· 61
　　三、精氨酸对炎症反应的影响 ································· 61
　　四、精氨酸对水体氨毒性的缓解作用 ································· 61
第五节　赖氨酸与精氨酸、谷氨酰胺的交互作用 ································· 62
第六节　小结与展望 ································· 62
参考文献 ································· 62

第七章　亮氨酸 ································· 67

第一节　亮氨酸的理化性质 ································· 67
第二节　水产动物亮氨酸需要量 ································· 67
第三节　亮氨酸与其他氨基酸的相互作用 ································· 69
　　一、亮氨酸与缬氨酸的相互作用 ································· 69
　　二、亮氨酸与异亮氨酸的相互作用 ································· 69
　　三、亮氨酸与其他氨基酸的作用关系 ································· 69
第四节　亮氨酸对水产动物蛋白质代谢的影响 ································· 70
第五节　亮氨酸对水产动物免疫功能的影响 ································· 70
第六节　亮氨酸对水产动物抗氧化功能的影响 ································· 71
第七节　亮氨酸对肠道发育的影响 ································· 71
第八节　小结与展望 ································· 72
参考文献 ································· 72

第八章　异亮氨酸 ································· 76

第一节　异亮氨酸的理化性质 ································· 76
第二节　水产动物异亮氨酸需要量 ································· 76

第三节　异亮氨酸与其他支链氨基酸的相互作用 …………………………………… 81
　第四节　异亮氨酸的生理功能 …………………………………………………………… 81
　　一、异亮氨酸对蛋白质代谢的影响 …………………………………………………… 81
　　二、异亮氨酸对机体免疫功能的影响 ………………………………………………… 82
　　三、异亮氨酸对机体抗氧化功能的影响 ……………………………………………… 82
　　四、异亮氨酸对肠道健康的影响 ……………………………………………………… 83
　　五、异亮氨酸对肉质的影响 …………………………………………………………… 84
　第五节　小结与展望 ……………………………………………………………………… 84
　参考文献 …………………………………………………………………………………… 84

第九章　缬氨酸 …………………………………………………………………………… 90

　第一节　缬氨酸的理化性质 ……………………………………………………………… 90
　第二节　水产动物缬氨酸需要量 ………………………………………………………… 90
　第三节　缬氨酸与支链氨基酸的相互作用 ……………………………………………… 93
　第四节　缬氨酸对水产动物蛋白质代谢的影响 ………………………………………… 94
　第五节　缬氨酸对水产动物免疫功能的影响 …………………………………………… 94
　第六节　缬氨酸对水产动物抗氧化能力的影响 ………………………………………… 95
　第七节　缬氨酸对肠道健康的影响 ……………………………………………………… 95
　第八节　小结与展望 ……………………………………………………………………… 96
　参考文献 …………………………………………………………………………………… 96

第十章　色氨酸 …………………………………………………………………………… 100

　第一节　色氨酸的理化性质 ……………………………………………………………… 100
　第二节　色氨酸的代谢途径 ……………………………………………………………… 100
　第三节　水产动物色氨酸需要量 ………………………………………………………… 102
　第四节　色氨酸对水产动物摄食量的影响 ……………………………………………… 103
　第五节　色氨酸对水产动物应激反应的影响 …………………………………………… 104
　第六节　色氨酸对水产动物抗病力、抗氧化和免疫功能的影响 ……………………… 105
　第七节　小结与展望 ……………………………………………………………………… 106
　参考文献 …………………………………………………………………………………… 106

第十一章　苯丙氨酸 ……………………………………………………………………… 111

　第一节　苯丙氨酸的理化性质 …………………………………………………………… 111
　第二节　苯丙氨酸的合成与代谢途径 …………………………………………………… 111
　　一、苯丙氨酸的生物合成 ……………………………………………………………… 111
　　二、苯丙氨酸的代谢 …………………………………………………………………… 112
　第三节　水产动物苯丙氨酸需要量 ……………………………………………………… 112
　第四节　苯丙氨酸与其他营养素的相互作用 …………………………………………… 115

第五节　苯丙氨酸对水产动物蛋白质代谢及健康的影响 ･･････････････････････････････････ 115
　　一、苯丙氨酸对水产动物蛋白质沉积率的影响 ･･････････････････････････････････････ 115
　　二、苯丙氨酸对水产动物体成分的影响 ･･ 116
　　三、苯丙氨酸对水产动物血清生化指标的影响 ･･････････････････････････････････････ 116
　　四、苯丙氨酸对水产动物肠道健康的影响 ･･ 117
第六节　小结与展望 ･･ 117
参考文献 ･･ 117

第十二章　组氨酸 ･･ 122

第一节　组氨酸的理化性质 ･･ 122
第二节　水产动物组氨酸需要量 ･･ 122
第三节　组氨酸对水产动物肉质的影响 ･･ 124
第四节　组氨酸对水产动物抗氧化能力的影响 ･･ 125
第五节　组氨酸对水产动物消化吸收能力的影响 ･･････････････････････････････････････ 125
第六节　组氨酸对水产动物免疫功能与健康的影响 ････････････････････････････････････ 126
第七节　组氨酸和其他氨基酸的相互作用 ･･ 126
第八节　小结与展望 ･･ 127
参考文献 ･･ 127

第三篇　其他氨基酸及肽营养

第十三章　牛磺酸 ･･ 135

第一节　牛磺酸的理化性质 ･･ 135
第二节　水产动物饲料牛磺酸的来源 ･･ 135
第三节　水产动物牛磺酸合成能力 ･･ 136
第四节　水产动物牛磺酸需要量 ･･ 137
第五节　牛磺酸对水产动物营养物质代谢的影响 ･･････････････････････････････････････ 139
　　一、牛磺酸对脂肪代谢的影响 ･･ 139
　　二、牛磺酸对糖代谢的影响 ･･ 140
　　三、牛磺酸对蛋白质代谢的影响 ･･ 140
第六节　牛磺酸对水产动物健康的影响 ･･ 140
　　一、牛磺酸对水产动物抗氧化能力及免疫力的影响 ････････････････････････････････ 140
　　二、牛磺酸对水产动物抗胁迫应激能力的影响 ････････････････････････････････････ 141
第七节　牛磺酸与其他营养物质的相互作用 ･･ 142
第八节　小结与展望 ･･ 143
参考文献 ･･ 143

第十四章　谷氨酰胺 ･･ 148

第一节　谷氨酰胺的理化性质 ･･ 148

第二节　谷氨酰胺的合成与代谢 …………………………………………………… 148
　　　一、谷氨酰胺的合成 ………………………………………………………………… 148
　　　二、谷氨酰胺的代谢 ………………………………………………………………… 149
　　第三节　谷氨酰胺的转化 ………………………………………………………… 149
　　第四节　谷氨酰胺对水产动物生长性能的影响 ………………………………… 149
　　第五节　谷氨酰胺对水产动物肠道发育与健康的影响 ………………………… 150
　　第六节　谷氨酰胺对水产动物抗氧化能力的影响 ……………………………… 151
　　第七节　谷氨酰胺对水产动物的其他影响 ……………………………………… 152
　　第八节　小结与展望 ……………………………………………………………… 152
　　参考文献 …………………………………………………………………………… 153

第十五章　多肽 …………………………………………………………………… 156

　　第一节　鱼源生物活性肽的分类 ………………………………………………… 156
　　　一、具有α-螺旋结构的鱼源生物活性肽 ………………………………………… 156
　　　二、含有二硫键的鱼源生物活性肽 ……………………………………………… 159
　　　三、组蛋白衍生肽 ………………………………………………………………… 161
　　第二节　鱼源生物活性肽的生物学活性与作用机理 …………………………… 162
　　　一、抗细菌 ………………………………………………………………………… 162
　　　二、抗病毒 ………………………………………………………………………… 162
　　　三、抗肿瘤 ………………………………………………………………………… 163
　　　四、其他活性 ……………………………………………………………………… 163
　　第三节　鱼源生物活性肽的潜在功能及应用 …………………………………… 163
　　　一、免疫原性药物的开发 ………………………………………………………… 163
　　　二、抗氧化剂的开发 ……………………………………………………………… 164
　　第四节　小结与展望 ……………………………………………………………… 164
　　参考文献 …………………………………………………………………………… 165

第四篇　氨基酸的应用与检测

第十六章　水产动物氨基酸的应用与检测 …………………………………… 173

　　第一节　氨基酸的平衡和互补作用 ……………………………………………… 173
　　　一、氨基酸的平衡 ………………………………………………………………… 173
　　　二、氨基酸的不平衡 ……………………………………………………………… 173
　　　三、氨基酸的互补 ………………………………………………………………… 173
　　　四、理想蛋白质 …………………………………………………………………… 173
　　第二节　水产动物氨基酸的有效性 ……………………………………………… 174
　　　一、晶体氨基酸的利用 …………………………………………………………… 174
　　　二、提高晶体氨基酸的利用 ……………………………………………………… 175

第三节　氨基酸之间的交互作用 …………………………………………………… 176
　一、协同作用 …………………………………………………………………… 176
　二、颉颃作用 …………………………………………………………………… 177
第四节　水产动物氨基酸与营养性疾病 …………………………………………… 177
　一、氨基酸的缺乏 ……………………………………………………………… 177
　二、氨基酸的中毒 ……………………………………………………………… 178
第五节　氨基酸的检测 ……………………………………………………………… 178
　参考文献 ………………………………………………………………………… 179

附录 …………………………………………………………………………………… 181

　附表一　水产动物的氨基酸营养需要 ………………………………………… 181
　附表二　饲料原料的氨基酸组成 ……………………………………………… 182

第一篇
氨基酸的概况及需要量研究方法

第一章 氨基酸的概况

第一节 氨基酸的结构

蛋白质是由氨基酸组成的,通过氨基酸种类、数量和结合方式不同组成不同的蛋白质。动物对蛋白质的需要实质上是对氨基酸的需要,蛋白质的营养实质为氨基酸的营养。自然界中,常见的组成植物和动物体的蛋白质有18种氨基酸和2种酰胺。氨基酸的通式为R—HCNH$_2$—COOH。基本结构是α-碳原子上连接一个羧基、一个氨基、一个氢原子和一个侧链R基团。根据R基团极性可以把氨基酸分为非极性R基团氨基酸、带电荷极性R基团氨基酸、碱性氨基酸和酸性氨基酸,氨基酸通常可以用三个字母或者一个字母的缩写形式来表示(表1-1)。

表1-1 氨基酸的分类、分子式与简写符号

分类与名称		分子式	英文名	简写符号		分子质量	含氮量(%)
非极性R基氨基酸	丙氨酸	C$_3$H$_7$O$_2$N	Alanine	Ala	A	89	15.7
	缬氨酸	C$_5$H$_{11}$O$_2$N	Valine	Val	V	117	12.0
	亮氨酸	C$_6$H$_{13}$O$_2$N	Leucine	Leu	L	131	10.7
	异亮氨酸	C$_6$H$_{13}$O$_2$N	Isoleucine	Ile	I	131	10.7
	脯氨酸	C$_6$H$_9$O$_2$N	Proline	Pro	P	115	12.2
	苯丙氨酸	C$_9$H$_{11}$O$_2$N	Phenylalanine	Phe	F	165	8.5
	色氨酸	C$_{11}$H$_{12}$O$_2$N$_2$	Tryptophan	Trp	W	204	13.7
	蛋氨酸	C$_5$H$_{11}$O$_2$NS	Methionine	Met	M	149	9.4
不带电荷的极性R基氨基酸	甘氨酸	C$_2$H$_5$O$_2$N	Glycine	Gly	G	75	18.7
	丝氨酸	C$_3$H$_7$O$_3$N	Serine	Ser	S	105	13.3
	苏氨酸	C$_4$H$_9$O$_3$N	Threonine	Thr	T	119	11.8
	半胱氨酸	C$_3$H$_7$O$_2$NS	Cysteine	Cys	C	121	11.6
	酪氨酸	C$_9$H$_{11}$O$_3$N	Tyrosine	Tyr	Y	181	7.7
	天冬酰胺	C$_4$H$_8$O$_3$N$_2$	Asparagine	Asn	N	132	21.2
	谷氨酰胺	C$_5$H$_{10}$O$_3$N$_2$	Glutamine	Gln	Q	146	19.2
带正电荷的R基氨基酸(碱性氨基酸)	赖氨酸	C$_6$H$_{14}$O$_2$N$_2$	Lysine	Lys	K	146	19.2
	精氨酸	C$_6$H$_{14}$O$_2$N$_4$	Arginine	Arg	R	174	32.2
	组氨酸	C$_6$H$_9$O$_2$N$_3$	Histidine	His	H	155	27.1

(续)

分类与名称		分子式	英文名	简写符号		分子质量	含氮量（%）
带负电荷的R基氨基酸（酸性氨基酸）	天冬氨酸	$C_4H_7O_4N$	Aspartic acid	Asp	D	133	10.5
	谷氨酸	$C_5H_9O_4N$	Glutamic acid	Glu	E	147	9.5

注：引自计成[1]。

α-氨基酸的氨基和羧基与同一碳原子相连，这在有机化学中被命名为 α-碳。不同氨基酸的差异在于 R 基团上，也就是通常所说的侧链上。R 基团的种类和大小可以从甘氨酸的一个氢原子到丙氨酸的一个甲基，再到色氨酸的大杂环结构。R 基团的结构差异造就了氨基酸的不同性质，包括分子质量、形状、电荷量及其他性质。氨基酸都有不对称碳原子（甘氨酸除外），具有 D-型和 L-型两种旋光异构体。自然界中最常见的氨基酸为 L-α-型，也是代谢中最重要的氨基酸。化学合成的蛋氨酸（即甲硫氨酸）多为 D、L 型混合物。D-型蛋氨酸可以通过异构酶转化为 L 型参与体内蛋白质的合成，二者具有相同的生物学效价。对于大多数氨基酸来说，由于缺乏相应的异构酶，D-型氨基酸不能被动物利用或利用率很低。目前有实验证据表明，无脊椎动物和包括硬骨鱼在内的低等脊椎动物能够更好地利用 D-型氨基酸[2,3]。特定的 D-氨基酸转氨酶（DAAO）可以催化 D-氨基酸脱羧成 α-酮酸，然后再与氨结合生成自然界广泛存在的 L-氨基酸。另一种情形是，消旋酶和差向异构酶能够将 D-异构体转化为外消旋的混合物，饲喂鱼类和无脊椎动物含有 D-氨基酸的饲料，似乎可以显著提高组织里这些酶的活性[3]。

氨基酸末端的 α-羧基能够和另一个氨基酸的 α-氨基通过共价键连在一起。通过这些连续的肽键，不同数目的氨基酸可以连在一起形成肽链。含有 2 个氨基酸的寡聚物被称为二肽，含有 2~20 个不等氨基酸的肽链被称为多肽。蛋白质分子通常含有 300 个左右的氨基酸，组成蛋白质分子的多肽为直链结构，不含有分支。氨基酸分子的结合顺序不同，就可以形成大量不同种类的蛋白质分子。蛋白质分子由其独特的氨基酸顺序所决定，而氨基酸序列则由生物体的遗传物质所编码。氨基酸序列是蛋白质的一级结构，肽键通过二硫键、氢键和范德华力的作用而形成二级、三级和四级结构。

此外，氨基酸具有两性电离特征，在不同 pH 溶液中可以解离为阳离子、阴离子或两性离子。

第二节 氨基酸的营养功能

蛋白质是组成机体结构物质（细胞）、体内代谢活性物质（激素、酶、免疫抗体）的主要成分，是组织更新、修补的原料。蛋白质占动物机体固形物总量的 50% 左右，肌肉、肝脏、脾脏、肾脏等器官的蛋白质含量可高达 80% 以上。氨基酸是构成蛋白质的基本物质，氨基酸通过不同的空间构型形成蛋白质，在体内发挥各种生理功能（表 1-2）。

除了作为蛋白质的组成部分之外，氨基酸在代谢中也具有很多功能。氨基酸对其他多种生物大分子非常重要。例如，氨基酸可参与形成辅酶分子，也可以作为铁红素、几丁质和嘌呤碱基等结构分子的生物合成前体；可以形成代谢的中间产物（如乙酸和丙酮酸），也可以形成神经递质、激素、生物胺和其他类的分子（如血清素、γ-羧基丁酸、褪黑素、一氧化氮和组胺，这些分子在生物体应对外界刺激时发挥作用）。然而，转化为其他生物活性分子的氨基酸数量，远少于蛋白质合成所需的氨基酸和发生降解反应的氨基酸。

表1-2 氨基酸的各种生物学功能

氨基酸	生物活性物质	功能
蛋氨酸	甲酰甲硫氨酰	蛋白质起始因子
	S-腺苷甲硫氨酸	甲基供体
	同型半胱氨酸	硫供体，维生素 B_{12} 营养的标识物
色氨酸	5-羟色胺（血管收缩素）	神经递质
	烟酰胺	B族维生素
酪氨酸	多巴胺	神经递质
	去甲肾上腺素	神经递质
	肾上腺素	激素
	甲状腺素	激素
精氨酸	一氧化氮	血管舒张；神经递质；雄性繁殖；胃肠蠕动
	多胺	调控RNA合成，维持膜稳定
组氨酸	组胺	强有力的血管舒张剂
谷氨酸	谷氨酰胺	嘌呤和嘧啶的合成
	谷胱甘肽	参与氧化还原反应
	γ-氨基丁酸（GABA）	神经递质
	能量	一些组织（黏膜）的能量来源
甘氨酸	卟啉类	参与血红蛋白构成
	嘌呤类	参与核酸构成
丝氨酸	鞘氨醇	膜组成成分
	半胱氨酸	对蛋白质作用非常重要
天冬氨酸	尿素、嘌呤、嘧啶	N供体
组氨酸	肌动蛋白和肌球蛋白	肌肉蛋白质降解标识物

注：引自刁其玉。[4]

第三节 氨基酸的代谢

一、氨基酸的吸收

在胃肠道中消化酶的作用下，摄取的蛋白质水解成为游离氨基酸、二肽和多肽。游离氨基酸穿过小肠中的黏膜细胞的纹状缘（或微绒毛）进入黏膜细胞内，再透过黏膜基底层的毛细血管进入血液系统。因此，小肠成为鱼类吸收氨基酸的关键场所。有实验表明，少量完整的蛋白质也可以被胃肠道吸收，这种吸收量非常少，但可能具有非常重要的生物学机制。例如，通过接收抗原来调节机体的免疫反应[5]。

游离氨基酸的转运主要是通过肠道黏膜上相应转运载体来完成。根据底物的类型，氨基酸转运系统分为酸性氨基酸转运系统、碱性氨基酸转运系统和中性氨基酸转运系统。根据转运过程是否依赖钠离子，可分为钠离子非依赖氨基酸转运系统和钠离子依赖氨基酸转运系统。大多数氨基酸的吸收是通过主动转运（依赖 Na^+ 和膜转运载体）进行，氨基酸转运系统主要有[6]：

1. 中性氨基酸转运系统 转运一氨基和一羧基氨基酸（丙氨酸、半胱氨酸、谷氨酰胺、天冬酰胺、组氨酸、异亮氨酸、亮氨酸、蛋氨酸、苯丙氨酸、丝氨酸、苏氨酸、色氨酸、酪氨酸、缬氨酸），为主动转运（Na^+ 依赖型），速度快，氨基酸之间存在竞争转运。

2. 碱性氨基酸转运系统 转运二氨基的氨基酸（精氨酸、赖氨酸、鸟氨酸和胱氨酸），为主动转运（Na^+ 依赖型），速度较快（约为中性氨基酸转运系统的10%）。

3. 酸性氨基酸转运系统 转运二羧基的氨基酸（天冬氨酸和谷氨酸），部分为 Na^+ 依赖型，可能为主动转运。因天冬氨酸和谷氨酸被肠黏膜细胞摄入后，迅速进行转氨反应而代谢，很难判断其吸收是否逆浓度梯度进行。

4. 亚氨基酸和甘氨酸转运系统 转运两种亚氨基酸（脯氨酸和羟脯氨酸）及甘氨酸，可能不需要 Na^+，其转运系统速度比其他三种系统低。

氨基酸的转运系统间可能存在交互作用，如一些中性氨基酸可通过碱性氨基酸转运系统吸收，中性氨基酸也可促进碱性氨基酸的吸收。同时，大量的游离氨基酸同时存在于肠道中，可能会导致运转机制的饱和竞争。相关实验表明，精氨酸和赖氨酸有共同的运输载体，且赖氨酸与载体的亲和力不如精氨酸的强。这说明，氨基酸的运转量与小肠黏膜表面通道或载体的亲和力及饱和性有关。不同载体系统对各种氨基酸的亲和力不同，各种氨基酸所共享的载体系统数目亦不同，这会出现尽管小肠游离氨基酸的整体浓度升高，但只表现出某些氨基酸的吸收速率提高的现象。若肠道内氨基酸浓度过高，因转运载体的过饱和而没有被吸收，最后只能从排泄系统中直接排出[7]。

二、氨基酸降解

细胞内蛋白质代谢是以氨基酸为中心的代谢。食物蛋白质经消化吸收的氨基酸（外源性氨基酸）与体内组织蛋白质分解产生的氨基酸（内源性氨基酸），加上一部分体内合成的非必需氨基酸，总称为氨基酸代谢库[8]。这个库里的氨基酸可以用来合成蛋白质，或者作为合成其他分子的前体。各种氨基酸具有共同的结构特点，故它们有共同的代谢途径，即脱氨作用和脱羧作用。其中脱氨作用是主要的代谢方式，脱羧作用是次要的代谢方式。

氨基酸的分解一般包括两步：第一步是脱氨，即将氨基酸的氨基脱去，成为氨或者谷氨酸的一个氨基。第二步是将脱氨后的碳骨架（α-酮酸）转化为三羧酸循环的中间产物[9]。氨基酸碳骨架所含有的能量可以在三羧酸循环（TCA 循环）中得以利用，或者被转化成脂肪酸和/或糖原供动物利用。每种氨基酸的碳骨架不同，因而不同的氨基酸通过各自的代谢途径进入到三羧酸循环中[2]。根据进入三羧酸循环的中间产物的类型，可以将氨基酸分成三种：生糖型、生酮型和糖酮兼生型。生糖型的氨基酸可以净生成丙酮酸或者α-酮戊二酸和草酰乙酸等三羧酸循环中间产物。这三种物质都可以通过糖异生途径合成葡萄糖。除赖氨酸和亮氨酸之外，其他氨基酸至少是部分生糖型氨基酸[2]。赖氨酸和亮氨酸是仅有的两种纯粹生酮型氨基酸，这两种氨基酸只能够生成乙酰辅酶 A 或乙酰乙酰辅酶 A。这两种产物都不能净生成葡萄糖，但却可以用于酮体的生成。它们的代谢途径的最终步骤非常相像，而且与脂肪酸的 β-氧化途径也相同[2]。异亮氨酸、苯丙氨酸、苏氨酸和酪氨酸既可以生成葡萄糖，也可以生成脂肪酸的前体，因此被列为糖酮兼生型脂肪酸[2]。

1. 氨基酸的脱氨作用

①氧化脱氨作用　氨基酸通过氨基酸氧化酶进行氧化脱氨作用，这一过程分两步：首先在氨基酸氧化酶的作用下，脱去一对氢原子，生成相对应的亚氨基酸，然后亚氨基酸自发水解生产相应的 α-酮酸，并释放出氨，总反应方程式为：

$$RCHNH_2COOH + 1/2 O_2 \longrightarrow RCOCOOH + NH_3$$

氨基酸氧化酶有两类，为 L-氨基酸氧化酶和 D-氨基酸氧化酶。L-氨基酸氧化酶在体内分布不广，活性也不高。D-氨基酸氧化酶在体内广泛存在，而且活性很强，但体内 D-氨基酸含量较少。D-氨基酸氧化酶能使 D-氨基酸氧化成 α-酮酸，再通过还原氨基化生产 L-氨基酸，这可以消除体内 D-氨基酸，防止非天然 D-氨基酸进入肽链[8]。

②转氨作用　一个氨基酸的氨基在转氨酶的作用下，转移到一个 α-酮酸分子上，氨基酸转变成 α-酮酸，而接收氨基的 α-酮酸则转变成氨基酸。转氨作用是氨基酸分解、合成及转变过程中的重要反应。多种氨基酸可以通过两种转氨酶偶联，进行连续转氨反应而参与分解代谢。转氨作用还可以将糖代谢产生的丙酮酸、α-酮戊二酸、草酰乙酸变为氨基酸，是沟通蛋白质代谢的桥梁[8]。

③联合脱氨作用　转氨作用中只有氨基的转移，而没有氨基的真正脱落，而且氧化脱氨作用仅有 L-谷氨酸脱氢酶的作用最强，其他氨基酸都不容易通过氧化脱氨作用进行脱氨。大量事实证明，体内氨基酸的脱氨作用方式是联合脱氨作用进行的[8]。

2. 氨基酸的脱羧作用　氨基酸在氨基酸脱羧酶的催化下进行脱羧反应，排出 CO_2 形成胺。氨基酸脱羧酶以磷酸吡哆醇为辅酶。脱羧作用是氨基酸分解的正常过程，但不是氨基酸分解的主要途径。某些氨基酸的脱羧产物具有很强的生理活性。如组氨酸、色氨酸、谷氨酸、酪氨酸和赖氨酸的脱羧产物分别为组胺、5-羟色胺、γ-氨基丁酸、儿茶酚胺和尸胺，它们都具有特殊的生理活性[8]。

$$RCHNH_2COOH \longrightarrow RCH_2NH_2 + CO_2$$

氨基酸的主要氧化终产物包括氨、二氧化碳和碳酸氢盐。其中，氨具有很强的毒性，为了控制其毒性，高等动物需要将氨转化为尿素，并通过尿排出体外。鱼类和甲壳类存在极为高效的转运机制，可以将氨通过鳃排出，因而不需要耗费能量将氨转化为尿素。绝大多数鱼、虾体内 80% 以上的含氮废物以氨的形式排出体外。鳃是主要的排泄器官，可将 75%~90% 的含氮废物排出体外[2,9]。

三、氨基酸的合成

不同生物合成氨基酸的能力不同，表现在合成氨基酸的原料和合成氨基酸的种类有很大差异，有的可以合成构成蛋白质的全部氨基酸，有的则不能合成全部。只有非必需氨基酸、半必需氨基酸、条件性必需氨基酸才能在体内合成。氨基酸的合成通过谷氨酸脱羧酶、转氨基作用和联合脱氨基作用的逆过程进行[6]。

1. 由 α-酮戊二酸形成的氨基酸 谷氨酸、谷氨酰胺、脯氨酸、精氨酸的碳架来自 α-酮戊二酸。

2. 由草酰乙酸形成的氨基酸 天冬氨酸、天冬酰胺的碳架来自草酰乙酸。

3. 由丙酮酸形成的氨基酸 丙氨酸的碳架来自丙酮酸。

4. 由 3-磷酸甘油酸形成的氨基酸 甘氨酸、丝氨酸和半胱氨酸的碳架来自 3-磷酸甘油酸。

5. 由磷酸烯醇式丙酮酸和 4-磷酸赤藓糖形成的氨基酸 酪氨酸的碳架来自磷酸烯醇式丙酮酸和 4-磷酸赤藓糖。

氨基酸被吸收以后，大部分用作蛋白质的合成，只有小部分多余的氨基酸会被脱去氨基，或是进入三羧酸循环被氧化释放出能量，或是转化成脂质和糖。在代谢过程中，必需氨基酸更多用于生长，合成蛋白质保留在体内，而非必需氨基酸更容易被氧化释放能量。

第四节 氨基酸的营养分类

从动物营养角度，将氨基酸分为必需氨基酸、半必需氨基酸、条件性必需氨基酸和非必需氨基酸。

1. 必需氨基酸 必需氨基酸是指动物体内不能合成或合成的数量不能满足需要，必须通过饲料来获取的氨基酸。研究表明，鱼类的必需氨基酸有 10 种，分别为异亮氨酸、亮氨酸、赖氨酸、蛋氨酸、苯丙氨酸、苏氨酸、色氨酸、缬氨酸、精氨酸和组氨酸[8]。

2. 半必需氨基酸 半必需氨基酸是指在一定条件下能代替或节省部分必需氨基酸需要的氨基酸，包括半胱氨酸或胱氨酸、酪氨酸及丝氨酸。苯丙氨酸和酪氨酸为芳香族氨基酸，苯丙氨酸是必需氨基酸，可以转化为酪氨酸，可以满足酪氨酸的需要，但酪氨酸不能转化为苯丙氨酸，因此饲料中必须保证足够的苯丙氨酸。同样，蛋氨酸在动物体内可以转化为半胱氨酸和胱氨酸，可满足半胱氨酸和胱氨酸的需要，半胱氨酸和胱氨酸可以节约蛋氨酸需要，但半胱氨酸和胱氨酸不能转化为蛋氨酸。

3. 条件性必需氨基酸 条件性必需氨基酸是指在特定情况下，必须由饲料提供的氨基酸，如精氨酸、组氨酸、脯氨酸和谷氨酰胺等。

4. 非必需氨基酸 非必需氨基酸是指动物体内能够合成满足需要，不需要由饲料提供的氨基酸。非必需氨基酸不是指动物在生长和维持生命的过程中不需要这些氨基酸，而是指当饲料中提供的非必需氨基酸不足时，动物体内可以合成这些氨基酸。

5. 限制性氨基酸 限制性氨基酸是指饲料中所含必需氨基酸的量与动物所需的必需氨基酸的量相比，比值偏低的氨基酸。由于这些氨基酸的不足，限制了氨基酸对其他氨基酸的利用。其中比值最低的称第一限制性氨基酸，以后依次为第二、第三限制性氨基酸……大多数植物蛋白源对水产动物来说，蛋氨酸和赖氨酸往往是限制性氨基酸[8]。

参考文献

[1] 计成. 动物营养学 [M]. 北京：高等教育出版社，2008.

[2] Cowey C B, Walton M J. Intermediary metabolism [C] // Halver J E. Fish nutrition. London: Academic Press, 1989.

[3] Deng J X, Zhang L, Tao H, et al. D‑lysine can be effectively utilized for growth by common carp (*Cyprinus carpio*) [J]. Aquaculture Nutrition, 2011, 17 (2): 467‑475.

[4] 刁其玉. 动物氨基酸营养与饲料 [M]. 北京: 化学工业出版社, 2007.

[5] McLean E, Rønsholdt B, Sten C, et al. Gastrointestinal delivery of peptide and protein drugs to aquacultured teleosts [J]. Aquaculture, 1999, 177 (1): 231‑247.

[6] 陈代文, 余冰. 动物营养学 [M]. 北京: 中国农业出版社, 2020.

[7] 霍湘, 王安利, 杨建梅. 鱼类对氨基酸的吸收代谢与需求 [J]. 水利渔业, 2005, 25 (3): 1‑3, 21.

[8] 麦康森. 水产动物营养与饲料学 [M]. 2版. 北京: 中国农业出版社, 2011.

[9] Kaushik S J, Cowey C B. Dietary factors affecting nitrogen excretion by fish [J]. Nutritional Strategies and Aquaculture Waster, 1990: 3‑20.

第二章 水产动物氨基酸需要量研究方法

水产动物对必需氨基酸定量需要研究方法有几种,但每一种方法都被认为有一些不足或缺陷。

一、生长试验法

绝大多数的必需氨基酸需要研究是通过传统的生长反应试验来完成的,这类试验在满足动物对所有其他营养素需要的基础上,设置缺乏一种必需氨基酸的基础饲料,然后添加不同浓度的该种氨基酸。该方法是由 Halver 及其同事在研究大鳞大麻哈鱼对氨基酸的需要时建立的,后来被许多学者所采用和发展。该方法使用纯化饲料、化学成分确定的饲料或天然饲料,用适宜的氨基酸混合饲料,将其中某一必需氨基酸以浓度梯度法变动其含量,制成试验饲料,进行生长试验;然后根据剂量-增重效应关系确定研究对象对氨基酸的需要量。除试验氨基酸外,其他必需氨基酸的组成模式与参照蛋白(如鸡全卵蛋白或鱼体蛋白)相一致[1]。

必需氨基酸的需要量的定量估算是基于剂量-反应曲线分析,通常增重是主要分析标准,取得最大增重的最低必需氨基酸含量即被定义为氨基酸最低需要量[2]。除了用生长作为反应参数外,越来越多的试验参考蛋白质和必需氨基酸沉积指标,在一些研究中,基于蛋白质和氨基酸沉积而估算的需要量往往高于以增重为标准而估算的需要量。

生长试验法是研究鱼、虾类营养素需要量的经典方法。对于大多数鱼、虾来说,在氨基酸组成相似的条件下,鱼类摄食添加游离氨基酸的纯化饲料,其生长率往往低于摄食由蛋白质提供氨基酸的饲料。因此,采用该方法测定的氨基酸需要量是在低于正常生长情况下测定的。另外,该方法还有工作量大、试验费用高、耗时多等缺点。

二、全鱼氨基酸组成

绝大多数摄食优质饲料的鱼类可将摄取的 25%～55%氨基酸沉积为体内的蛋白质。因而,蛋白质的沉积是鱼类氨基酸需要的一个主要决定因素。此外,沉积体蛋白的组成也可为饲料中动物必需氨基酸的组成提供借鉴。早期的鱼类氨基酸需要是基于鱼、卵或者全鱼蛋白的氨基酸组成来估计的。在绝大多数情况下,体内必需氨基酸含量和必需氨基酸需要量(占蛋白质的百分比)之间存在着较好的一致性。根据全鱼或肌肉氨基酸组成以及必需氨基酸比例(必需氨基酸含量/全部必需氨基酸×1 000)[3],也可以对鱼类的必需氨基酸需要量进行简单的定量估算。这是基于"理想蛋白"模式。但基于鱼体蛋白估算的必需氨基酸需要量会高估亮氨酸和赖氨酸等优先在体蛋白中沉积的氨基酸需要量，却低估蛋氨酸、苏氨酸、组氨酸和精氨酸等在代谢中起重要作用的氨基酸需要量[2]。

该方法至今仍被用来初步估计鱼、虾类氨基酸需要量，尤其是当营养需要研究还很缺乏的时候。然而，由于被保留的氨基酸通常只占被消化氨基酸的一小部分（通常小于50%），因而沉积的必需氨基酸和动物的必需氨基酸需要量并不完全一致，某些氨基酸只限在组织内沉积，而某些氨基酸具有更加活跃的代谢作用，从而具有较低的沉积效率。

三、氨基酸需要量的析因方法

析因模型近年来开始在鱼类上使用[4]。在经常使用的模型中，必需氨基酸的利用被表示为基础需要量、过量必需氨基酸的分解代谢、不可避免的分解代谢、用于供能的优先分解代谢和体蛋白的沉积[5]。这种方法在鱼类中应用时存在很多局限性，但却为估算鱼类的必需氨基酸需要量提供了相对直接的方法。

1. 合成鱼体蛋白质需要 氨基酸在鱼体内的主要代谢途径即合成蛋白质。Fauconneau等[6]利用同位素标记的方法标记亮氨酸后注射到虹鳟体内，研究发现进入到氨基酸库的亮氨酸有60%用来合成蛋白质。由于鱼体的生长和鱼体内的蛋白质沉积有很大的相关性，因此鱼体的生长和氨基酸需要量也有很大的相关性。

2. 维持鱼体内氨基酸的需要量 维持鱼体内氨基酸的需要量主要是指氨基酸的总量能够维持体内氨基酸代谢库平衡的需要。维持体内氨基酸的平衡包括体内肠道蛋白降解、黏蛋白以及其他物质的分泌造成的氨基酸的流失。氨基酸也需要作为许多代谢物质、神经递质、辅酶等的前体。

研究鱼类维持基础代谢的氨基酸的需要量一般是通过线性回归方程得到的。Rodehutscord等[7]研究了虹鳟（50 g）用于维持基础代谢的氨基酸的需要量［赖氨酸4 mg/(kg·d)，色氨酸2 mg/(kg·d)，组氨酸2 mg/(kg·d)］。Bodin等[8]的研究结果显示虹鳟维持赖氨酸需要量要更高一些［赖氨酸24 mg/(kg·d)］。Rollin等[9]的研究结果表明大西洋鲑幼鱼（1~2 g）的苏氨酸维持需要量为5~7 mg/(kg·d)。这些研究结果表明鱼类用于维持基础代谢的最低氨基酸需要量与组织中氨基酸组成有很大差异，不同鱼种间维持的氨基酸需要量也存在较大差异。

3. 不可避免的氨基酸分解代谢 不可避免的氨基酸分解代谢是指在能量供应不影响蛋白质合成的前提下，通过主动降解途径分解氨基酸的过程。这种降解似乎是由于细胞内存在无法完全停止的降解机制，从而不可避免地发生。因此，即使当氨基酸的摄取量低于最大蛋白质沉积（Maximal protein deposition rate，PD_{max}）所需的氨基酸量，吸收的部分氨基酸仍会被降解掉。当摄取的氨基酸不足以满足动物最大蛋白沉积时，不可避免的氨基酸分解代谢就影响着蛋白质沉积过程中氨基酸利用的边际效率。但氨基酸摄取量达到PD_{max}所需要的70%~100%时，不可避免的氨基酸分解代谢的速率（以摄取的可利用氨基酸比例来表示）被认为保持不变。在满足基础氨基酸需要后，动物体通过不可避免的降解途径消耗掉20%~40%的可消化氨基酸[2]。

4. 优先的分解代谢 优先分解代谢是指当饲料能量成为蛋白质沉积的限制因素时，氨基酸通过降解来提供能量，这意味着动物将必需氨基酸从蛋白质合成中分配出来。通过降解途径来满足特定的代谢需要[5]。优先分解代谢和不可避免的氨基酸分解代谢的区别可以从代谢能摄入和蛋白质沉积之间的关系的斜率中推断出来[10]。

在很大程度上，鱼类似乎可以通过氨基酸降解途径来满足总能量（三磷酸腺苷，ATP）的需求。鱼类的优先分解代谢与不可避免的氨基酸分解代谢或许很难准确区分开来。在代谢能和净能摄入充足的情况下，氨基酸降解与不可避免的氨基酸损失（基础需求与不可避免的氨基酸降解）或过量氨基酸的降解抑或氨基酸的优先降解（作为能源）之间的相关程度，至今仍很难界定。

5. 过量氨基酸的分解代谢　当摄入的氨基酸供给超过蛋白质沉积、基础需求、不可避免的氨基酸分解以及优先分解代谢所需要氨基酸时，过量氨基酸将会被降解掉。但饲料出现一种或多种氨基酸缺乏时（相当于营养需要），蛋白质沉积和其他氨基酸的保留都会受到抑制，而脱氨作用和分解代谢则被激活。

析因法是建立在鱼体生长模型的基础上来计算用于维持鱼体最大生长的饲料中氨基酸的需要量，常用公式 $C=(G+E)/(A/F)$ 来表示。其中，C 表示氨基酸在饲料中的水平；A 表示某种氨基酸在鱼体的可利用水平；F 代表摄食量；G 表示用于生长的氨基酸需要量；E 表示用于维持的氨基酸需要量[11]。

析因模型可以基于饲料成分、生长阶段、生长速度来估算必需氨基酸的需要量，因而具有非常实用的价值。目前，该模型主要的局限性在于：为每种必需氨基酸进行独立的估算，并假定必需氨基酸之间以及与其他营养素（如脂肪和葡萄糖）之间不存在相互作用，同时假定氨基酸利用效率完全不受生理状况影响。该模型还假定"摄食量""能量需求"和"饲料系数"不是"决定因素"，也就是说，它们与饲料组成、营养素摄取、营养素之间的相互作用、靶蛋白的改变、脂肪沉积或者生理状态等因素"相互独立"。

四、血液和肌肉中的氨基酸浓度

当摄取的氨基酸不能达到动物氨基酸需要量的时候，血清和组织中检测的必需氨基酸浓度会维持在较低水平；当必需氨基酸的需要量得到满足并超过所需要的氨基酸时，血清和组织内的必需氨基酸才会上升到较高浓度。该方法在验证必需氨基酸需要量时非常有用[2]。如在斑点叉尾鮰中，以血清赖氨酸数据[12]验证通过增重率获得的赖氨酸需要量，发现具有很好的应用性。然而，采用该技术估算必需氨基酸的需要量并不总是完全可靠[9]。该技术的可靠性与所检测的必需氨基酸的性质、氨基酸间的相互作用及摄食与取血时间间隔有关[13]。

五、氨基酸氧化研究

基于组织中游离氨基酸浓度所做的直接或者间接氧化试验，为测定必需氨基酸在蛋白质合成和氧化之间的分配提供了方法[14]。直接氧化法是基于这样的一个原则：在限制性状态下，由于游离氨基酸库的浓度较低，将导致必需氨基酸的氧化速度缓慢；绝大多数氨基酸被用来合成蛋白，而被氧化的氨基酸量则非常低。因此，所研究的氨基酸氧化率非常低，直到需要量被满足后才迅速升高。当被研究的氨基酸摄取量使得氨基酸氧化迅速增加时，该摄取量就是此种氨基酸需要量的一个直接指标。间接氧化法是通过目测氨基酸之外的某种必需氨基酸（待测必需氨基酸）的氧化过程来实现的。在这套方法中，目标氨基酸的限制就会影响待测必需氨基酸在组织蛋白中的沉积，其造成的结果是待测必需氨基酸出现快速氧化。当饲料中目标氨基酸的含量增加时，组织蛋白的合成会持续增加，而被氧化的待测必需氨基酸也就成比例降低。产生上述现象的原因在于，待测必需氨基酸用于蛋白合成的量得到了相应的增加[14]。

参考文献

[1] 麦康森. 水产动物营养与饲料学 [M]. 2版. 北京：中国农业出版社，2011.

[2] National Research Council. Nutrient Requirements of Fish and Shrimp [M]. Washington, DC: National Academy Press, 2011.

[3] Arai. A purified test diet for Coho salmon, *Oncorhynchus kisutch*, fry [J]. Bulletin of the Japanese Society of Scientific Fisheries, 1981, 47: 547-550.

[4] Hauler R C, Carter C G. Re-evaluation of the quantitative dietary lysine requirement of fish [J]. Reviews in Fisheries Science, 2001, 9: 133-163.

[5] Moughan P J. Simulating the partitioning of dietary amino acids: new directions [J]. Journal of Animal Science, 2003, 81: 60-67.

[6] Fauconneau B, Tesseraud S. Measurement of leucine flux in rainbow trout (*Salmo gairdheri* R.) using osmotic pump. Preliminary investigations on influence of diet [J]. Fish Physiology and Biochemistry, 1990, 8: 24-44.

[7] Rodehutscord M A, Becker M P, Pfeffer E. Response of rainbow trout (*Oncorhynchus mykiss*) to supplements of individual essential amino acids in a semipurified diet, including an estimate of the maintenance requirement for essential amino acids [J]. Journal of Nutrition, 1997, 127: 1163-1175.

[8] Bodin N B, Govaerts T, Abboudi C D, et al. Protein level affects the relative lysine requirement of growing rainbow trout (*Oncorhynchus mykiss*) fry [J]. Brit. J. Nutr, 2009, 102: 27-53.

[9] Rollin X M, Mambrini T, Abboudi Y, et al. The optimum dietary indispensable amino acid pattern for growing Atlantic Salmon (*Salmo salar* L.) fry [J]. British Journal of Nutrition, 2003, 102: 234-245.

[10] Mohn S, de Lange C F M. The effect of body weight on the upper protein limit to protein deposition in a defined population of growing gilts [J]. Journal of Animal Science, 1998, 76: 124-133.

[11] 高煜霞, 田丽霞, 刘永坚. 鱼类氨基酸研究进展 [J]. 广东饲料, 2012, 21 (增刊1): 53-57.

[12] Wilson R P, Harding D E, Garling Jr. Effect of dietary pH on amino acid utilization and lysine requirement of fingerling channel catfish [J]. Journal of Nutrition, 1977, 107: 166-170.

[13] Mambrini M, Kaushik S. Indispensable amino acid requirements of fish: correspondence between quantitative data and amino acid profiles of tissue protein [J]. Journal of Applied Ichthyology, 1995, 11: 240-247.

[14] Kim H G, McMillam I, Bayley H. Determination of amino acid requirement of young pigs using an indicator animo acid [J]. British Journal of Nutrition, 1983, 50: 369-382.

第二篇
水产动物必需氨基酸营养

第三章 赖氨酸

为探讨在植物蛋白原料被高比例应用于水产动物饲料配方的情况下,赖氨酸作为饲料中第一或者第二限制性氨基酸的精准营养需要,本章系统梳理并综述了国内外近二十年水产动物赖氨酸需要量研究成果,针对现有的相关研究成果,阐述了影响赖氨酸需要量的相关因素,并介绍了赖氨酸对水产动物消化吸收、免疫功能、氮代谢等方面的影响。

第一节 赖氨酸的理化性质

赖氨酸(Lysine,Lys),分子式 $C_6H_{14}N_2O_2$,分子质量 146.19,熔点 215 ℃,分子结构如图 3-1 所示,易溶于水,微溶于醇,不溶于醚。按光学活性分,赖氨酸有 L 型(左旋)、D 型(右旋)和 DL 型(消旋)3 种构型。只有 L 型才能为生物所利用。

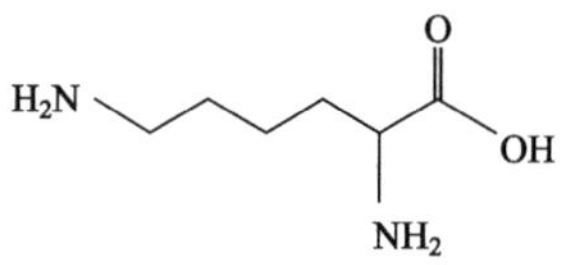

图 3-1 赖氨酸分子结构

第二节 水产动物赖氨酸需要量

表 3-1 总结了近二十年来关于水产动物赖氨酸需要量的研究报道。对大多数鱼类来说,赖氨酸需要量都在 1.34%～3.83%(占饲料)或 2.94%～8.64%(占饲料蛋白)的范围内;对于虾类来说,赖氨酸需要量都在 0.77%～2.63%(占饲料)或 2.02%～5.84%(占饲料蛋白)的范围内;对于蟹类来说,赖氨酸需要量都在 2.34%～2.41%(占饲料)或 4.87%～6.39%(占饲料蛋白)的范围内。

表 3-1 水产动物赖氨酸需要量

	品种	体重	模型	指标	需要量(%) 占饲料干物质	需要量(%) 占饲料蛋白	参考文献
海水鱼类	红鳍东方鲀 Takifugu rubripes	13.83 g	二次多项式模型	特定生长率	3.83	7.66	张庆功等[1]

(续)

品种		体重	模型	指标	需要量（%）		参考文献
					占饲料干物质	占饲料蛋白	
海水鱼类	卵形鲳鲹 *Trachinotus blochii*	6.28 g	二次多项式模型	增重率	2.40	5.71	Ebeneezar 等[2]
				饲料系数	2.42	5.76	
				特定生长率	2.45	5.83	
				蛋白质效率	2.42	5.76	
		14.78 g	折线模型	增重率	2.94	6.70	杜强等[3]
	大黄鱼 *Pseudosciaena crocea* Richardson	2.75 mg	二次多项式模型	特定生长率	3.37	6.65	Xie 等[4]
		1.23 g	折线模型	定生长率	2.48	5.77	Zhang 等[5]
				饲料系数	2.45	5.70	
				蛋白质效率	2.43	5.65	
	半滑舌鳎 *Cynoglossus semilaevis* Günther	3.66 g	折线模型	特定生长率	2.85	5.78	代伟伟[6]
				蛋白滞留率	2.75	5.58	
		113.01 g		特定生长率	2.49	5.06	
				蛋白滞留率	2.43	4.92	
		347.30 g		特定生长率	2.46	4.99	
				蛋白滞留率	2.58	5.24	
	斜带石斑鱼 *Epinephelus coioides*	15.84 g	折线模型	增重率	2.83	5.56	Luo 等[7]
		22.07 g	二次多项式模型	特定生长率	3.04	6.63	王学武[8]
		102.51 g			2.61	5.80	
	印度囊鳃鲇 *Heteropneustes fossilis*	4.78 g	指数回归分析	赖氨酸保留效率	1.32	2.94	Khan 和 Abidi[9]
				蛋白质保留效率	1.41	3.14	
		6.96 g	二次多项式模型	增重率	2.19	5.76	Farhat 和 Khan[10]
				饲料系数	2.29	6.02	
			折线模型	增重率	2.0	5.26	
				饲料系数	1.98	5.21	
	银鲈 *Bidyanus bidyanus*	22.84 g	折线模型	增重率	2.06	5.30	Yang 等[11]
			二次多项式模型		2.32	5.96	
	大西洋鳕 *Gadus morhua*	100 g	指数回归模型	氮沉积率	2.38	4.76	Gridale-Helland 等[12]
				赖氨酸获得率	2.62	5.24	
	大西洋鲑 *Salmo salar*	62.8 g	指数回归模型	赖氨酸获得率	2.67	5.34	Gridale-Helland 等[13]
	六指马鲅 *Polydactylus sex filis*	3.0 g	折线模型	增重率	1.79	5.1	Dominy 等[14]

(续)

	品种	体重	模型	指标	需要量（%）		参考文献
					占饲料干物质	占饲料蛋白	
海水鱼类	黑鲷 *Sparus macrocephalus*	9.13 g	二次多项式模型	特定生长率	3.31	8.64	Zhou 等[15]
	军曹鱼 *Rachycentron canadum*	1.25 g	折线模型	特定生长率	2.33	5.30	Zhou 等[16]
		14.7 g	二次多项式模型	增重率	2.78	6.06	周萌等[17]
	鲈 *Lateolabrax japonicus*	5.50 g	折线模型	特定生长率	2.49	5.80	Mai 等[18]
				饲料系数	2.61	6.08	
				蛋白质效率	2.60	6.07	
		215.00 g	折线模型	特定生长率	2.68	6.09	门珂珂[19]
				饲料系数	2.47	5.61	
	许氏平鲉 *Sebastes schlegeli*	28.42 g	二次多项式模型	增重率	2.99	6.16	严全根等[20]
	金头鲷 *Sparus aurata L.*	3.5 g	折线模型	日蛋白沉积率	2.17	5.27	Marcouli 等[21]
				增重率	2.00	5.04	
	尖吻鲈 *Lates calcarifer*	13.12 g	折线模型	增重率	2.06	4.50	Murillo-Gurrea 等[22]
	条纹鲈 *Morone saxatilis*	6.0 g	最小二乘回归	增重率	1.98	4.71	Small 和 Soares[23]
				特定生长率	2.17	5.17	
				饲料系数	2.37	5.64	
淡水鱼类	虹鳟 *Oncorhynchus mykiss*	12.2 g	折线模型	增重率	2.23	5.01	Lee 等[24]
				特定生长率	2.75	6.18	
	吉富罗非鱼 *Oreochromis miloticus*	5.2 g	折线模型	增重率	1.80	5.62	Prabu 等[25]
				氮沉积率	1.83	6.13	
			二次多项式模型	增重率	1.93	5.72	
				氮沉积率	1.95	6.09	
	乌鳢 *Channa argus*	5.96 g	折线模型	特定生长率	2.87	6.65	尹东鹏等[26]
	日本银鲴 *Argyrosomus japonicus*	4.5 g	折线模型	特定生长率	3.17	7.35	Adesola 等[27]
	雅罗鱼 *Leuciscus*	13.44 g	二次多项式模型	增重率	3.26	8.81	左亚男等[28]
	尼罗罗非鱼 *Oreochromis niloticus L.*	274. g	二次多项式模型	出肉率	1.46	5.80	Michelato 等[29]
	团头鲂 *Megalobrama amblycephala*	3.34 g	折线模型	特定增重率	2.36	6.96	廖英杰等[30]
				蛋白沉积率	2.22	6.53	
		52.49 g	二次多项式模型	特定生长率	2.07	6.27	宋长友等[31]
		101.85 g	二次多项式模型	特定生长率	2.19	6.63	

（续）

	品种	体重	模型	指标	需要量（%）占饲料干物质	需要量（%）占饲料蛋白	参考文献
淡水鱼类	乌苏里拟鲿 *Pseudobagrus ussuriensis*	0.6 g	折线模型	增重率	3.38	8.31	王裕玉[32]
				蛋白质效率	3.91	9.61	
	鳡 *Elopichthys bambusa*	2.36 g	折线模型	特定生长率	2.90	6.30	杨威[33]
			二次多项式模型		3.41	7.41	
	胭脂鱼 *Myxocyprinus asiaticus*	1.81 g	折线模型	增重率	2.43	5.52	Lin 等[34]
				特定生长率	2.42	5.50	
				蛋白质效率	2.40	5.45	
	黄颡鱼 *Pelteobagrus fulvidraco*	1.48 g	折线模型	特定生长率	3.31	8.32	Cao 等[35]
		2.00 g	折线模型	增重率	2.61	5.80	邱红 等[36]
	锯腹脂鲤 *Piaractus mesopotamicus*	8.66 g	折线模型	增重率	1.64	7.13	Abimorad 等[37]
		4.3 g	二次多项式模型	增重率	1.45	4.44	Bicudo 等[38]
				饲料效率	1.51	4.72	
	建鲤 *Cyprinus carpio* var. *jian*	7.89 g	折线模型	增重率	1.88	5.90	Zhou 等[39]
	异育银鲫 *Carassius auratus gibelio*	7.85 g	折线模型	特定生长率	3.27	8.52	周贤君 等[40]
		52.53 g	二次多项式模型	特定生长率	1.78	5.14	王鑫 等[41]
				蛋白质沉积率	1.43	4.14	
	草鱼 *Ctenopharyngodon idella*	3.15 g	二次多项式模型	增重率	2.24	5.89	Wang 等[42]
	麦瑞加拉鲮 *Cirrhinus mrigala*	0.63 g	二次多项式模型	增重率	2.80	5.75	Ahmed 和 Khan[43]
	胡子鲇 *Mystus nemurus*	150 g	二次多项式模型	特定生长率	1.21	3.47	Tantikitti 和 Chimsung[44]
甲壳类	大西洋沟虾 *Palaemonetes varians*	17 mg	折线模型	增重率	2.42	5.38	Palma 等[45]
			指数回归分析		2.63	5.84	
	日本沼虾 *Macrobrachium nipponense*	0.13 g	折线模型	增重率	1.91	5.41	孙丽慧 等[46]
	凡纳滨对虾 *Litopenaeus vannamei*	0.54 g	折线模型	特定生长率	2.05	3.95	Xie 等[47]
		0.52 g	折线模型	特定生长率	2.05	4.92	曾雯娉[48]
	克氏原螯虾 *Procambarus clarkii*	7.60 g	二次多项式模型	增重率	1.66	5.87	张微微[49]
	斑节对虾 *Penaeus monodon*	2.4 g	Logistic 模型	氮沉积	0.77	2.02	Richard 等[50]

(续)

	品种	体重	模型	指标	需要量（%）		参考文献
					占饲料干物质	占饲料蛋白	
甲壳类	三疣梭子蟹 *Portunus trituberculatus*	7.86 g	折线模型	特定生长率	2.41	4.87	金敏[51]
	中华绒螯蟹 *Eriocheir sinensis*	2.03 g	二次多项式模型	增重率	2.34	6.39	叶金云等[52]

第三节 水产动物赖氨酸需要量及利用的影响因素

一、品种、食性、生长阶段

从表 3-1 总结的研究结果来看，肉食性海水鱼的赖氨酸需要量在 1.34%～3.83%（占饲料）或 2.94%～8.64%（占饲料蛋白）；淡水鱼类对赖氨酸的需要量在 1.46%～3.47%（占饲料）或 3.47%～9.61%（占饲料蛋白）。上述研究说明不同种类的鱼类对赖氨酸的最适需要量存在着明显的差异。

Xie 等[4]和 Zhang 等[5]分别发现初重为 2.75 mg 和 1.23 g 的大黄鱼对赖氨酸的最适需要量分别为占饲料的 3.37% 和占饲料的 2.43%～2.45%；代伟伟[6]发现初重为 3.66 g、113.0 g、347.30 g 的半滑舌鳎对赖氨酸需要量分别为占饲料的 2.75%～2.85%、2.43%～2.49% 和 2.46%～2.58%；Mai 等[18]和门珂珂[19]分别发现初重为 5.50 g 和 215.00 g 的花鲈对赖氨酸的最适需要量分别为占饲料的 2.49%～2.61% 和占饲料的 2.47%～2.68%；初重分别为 3.34 g、52.49 g 和 101.85 g 的团头鲂对赖氨酸需要量分别为占饲料的 2.22%～2.36%[29]、2.07% 和 2.19%[30]。说明对于同一种鱼类，不同的养殖规格及年龄，对赖氨酸的最适需要量也有所不同。低龄鱼和生长速度快的鱼类与体成熟和生长速度慢的鱼类相比需要更多的赖氨酸用于蛋白质的合成和沉积。

二、饲料因素

饲料的蛋白源组成会影响水产动物对赖氨酸的需要及利用能力。相较于动物性蛋白源，植物性蛋白源往往缺乏如赖氨酸和蛋氨酸等必需氨基酸[4]，因此当植物性蛋白作为饲料的主要蛋白源时，评估出的赖氨酸需要量往往高于以鱼粉作为饲料主要蛋白源时的需要量[53]。付锦锋[54]研究发现在虹鳟饲料中添加赖氨酸，当以植物蛋白作为基础蛋白质来源，外源赖氨酸的添加量为 0.8% 时（饲料赖氨酸水平为 2.1%），虹鳟的生长率与对照组（42% 蛋白组：以鱼粉为主要蛋白源，饲料赖氨酸水平为 2.0%）相比无显著差异，且排入水中的总氨氮和可溶性磷分别减少了 28% 和 71%。Thu 等[55]研究了在以小麦面筋、玉米面筋或芝麻油饼和小麦面筋的混合物为主要植物蛋白的饲料中添加赖氨酸对虹鳟赖氨酸保留率的影响，线性回归方程表明，虹鳟沉积 1 g 蛋白需要分别摄入 114、133 和 138 mg 的赖氨酸，小麦面筋组对赖氨酸保留率显著高于另外两组，表明在设计以植物蛋白源为主要蛋白源的饲料配方时应考虑赖氨酸利用效率的变化。

赖氨酸与同为碱性氨基酸的精氨酸存在颉颃作用，这一现象已在家禽[56]及小鼠[57]的实验中得到证实。因此，在设计饲料配方时应考虑赖氨酸与精氨酸的配比问题。但目前水产动物的营养学研究中赖氨酸与精氨酸间是否存在颉颃作用仍有争议。早期的相关研究表明在斑点叉尾鮰（*Ictalurus punctatus*）[58]、虹鳟[59]和黄鲈（*Perca flavescens*）[60]上并未出现赖氨酸与精氨酸的配比不同影响鱼类生长及饲料利用率。而最近的研究发现精氨酸和赖氨酸的交互作用对大菱鲆（*Scophthatmus maximus* L.）幼鱼的饲料效率、全鱼及肌肉粗蛋白含量、肌肉氨基酸组成产生了显著影响。与精氨酸相比，赖氨酸为主要影响因素，适量添加赖氨酸可以促进生长，而添加量过高赖氨酸会与精氨酸产生颉颃作用，抑制生长、饲料利用和肌肉氨基酸沉积[61]。在对大西洋鲑的研究中发现过高的赖氨酸添加量会限制肌肉中精氨酸的利用效率[62]。Dong 等[63]对罗氏沼虾（*Macrobrachium rosenbergii*）的研究发现，当以血清中游离精氨酸含量和精氨酸与赖氨酸的保留率作为评价指标时，精氨酸和赖氨酸表现出明显的颉颃作用。因此，在水产动物中可能存在更复杂的赖氨酸及精氨酸代谢及调控机制，简单地遵循哺乳动物和鸟类的结论是不合理的。目前在水产动物上，虽然有大量关于赖氨酸与精氨酸配比的研究，但结果不尽相同，品种、养殖环境、饲料配方等均会影响水产动物对氨基酸的需要及利用能力。因此，在未来的研究中通过深入探讨赖氨酸与精氨酸在代谢中的相互关系，进而确定饲料配方中二者的适宜比例，为养殖实践奠定理论基础。

三、赖氨酸的添加形式

为了平衡饲料的氨基酸组成，满足水产动物对饲料中赖氨酸的需要，最好的方式是在饲料中添加晶体赖氨酸。晶体赖氨酸存在两个异构体，即 D-赖氨酸和 L-赖氨酸。Friedman 和 Gumbmmann[64]发现在小鼠饲料中添加 D-赖氨酸并不能促进机体正常生长及能量代谢。但 Deng 等[65]对鲤的研究结果表明，在以豆粕为主要蛋白源的饲料中添加 0.4% D-赖氨酸和 L-赖氨酸都能够增强生长性能及营养物质的利用能力，而与赖氨酸代谢相关的 D-氨基酸氧化酶和 D-天冬氨酸氧化酶活性均显著高于 L-赖氨酸组。关于水产动物对 D-赖氨酸和 L-赖氨酸的代谢机制有待进一步研究。

饲料级商品赖氨酸的主要形式包括两种：赖氨酸盐酸盐（L-Lys·HCl）和赖氨酸硫酸盐（L-Lys·H_2SO_4）。L-Lys·HCl 常用的为 98% L-Lys·HCl（简称 98-Lys，Lys 含量≥78%）；L-Lys·H_2SO_4 常用的有 Biolys60（Lys 含量≥46.8%）、BiolysTM（Lys 含量≥50.7%）、65% L-Lys·H_2SO_4（Lys 含量≥51%）及 70% L-Lys·H_2SO_4（Lys 含量≥55%）。水产动物对于 L-Lys·HCl 和 L-Lys·H_2SO_4 的利用效果存在一定差异。L-Lys·HCl 和 Lys·H_2SO_4 在斑点叉尾鮰[66]、虹鳟[67]上的生物学效价相当。胡凯等[68]对生长中期草鱼的研究发现，饲料中添加适宜水平的 L-Lys·H_2SO_4 使饲料赖氨酸含量达到 1.2%时显著提高生长中期草鱼的增重率，特定生长率，采食量，全肠脂肪酶、淀粉酶活力，肝胰脏谷草转氨酶和谷丙转氨酶活力，肝体指数与肠体指数以及前、后肠皱襞高度；显著降低血清谷草转氨酶和谷丙转氨酶活力，且 L-Lys·H_2SO_4 对上述指标的作用效果显著优于 L-Lys·HCl，这表明与 L-Lys·HCl 相比，L-Lys·H_2SO_4 能更有效地提高生长中期草鱼的消化吸收能力，进而促进其生长。在斑节对虾饲料中分别添加 L-Lys·HCl 和 L-Lys·H₂SO₄ 后，前者能显著提升斑节对虾的摄食量，使其获得更高的增重率和饲料效率[69]。赵金鑫等[70]发现，在斑点叉尾鲴饲料中添加 L-Lys·HCl 和 L-Lys·H₂SO₄ 投喂 60 d 后，L-Lys·H₂SO₄ 组的增重率为 L-Lys·HCl 组的 104%，即 L-Lys·H₂SO₄ 的生物学效价为 L-Lys·HCl 的 1.04 倍。需要指出的是，L-Lys·HCl 和 L-Lys·H₂SO₄ 均为晶体氨基酸，虽与普通晶体氨基酸相比能被机体更好地利用，但仍存在与结合态氨基酸无法同步吸收的问题，因此斑点叉尾鲴对二者利用效果仍不理想。

晶体氨基酸的水溶失率高是影响水产动物利用饲料外源晶体氨基酸的第一个主要因素。目前在水产饲料行业中主要通过包被或者微囊化的缓释处理达到降低晶体氨基酸的水溶失率的目的。刘永坚等[71]发现在 10 min 的溶失率实验后，包膜赖氨酸组在水中 4.81% 的溶失率显著低于晶体赖氨酸组的 13.22%。Alam 等[72]以饲料中添加 1.21% 晶体蛋氨酸和 1.45% 晶体赖氨酸为对照组，以仅添加 1.45% 的包被赖氨酸、1.21% 包被蛋氨酸和添加 1.21% 包被蛋氨酸和 1.45% 包被赖氨酸为试验组，在 15 min 的溶失率实验后发现，包膜赖氨酸组及包被赖氨酸和包被蛋氨酸组的赖氨酸溶失率仅占饲料的 0.1% 和 0.08%，而晶体赖氨酸组却高达 0.138%。王冠等[73]发现硬化油脂包膜赖氨酸酸的水中溶失率（14.3%±0.7%）仅相当于晶体氨基酸溶失率（57%±2.1%）的 25.1%。钱前[74]测定了晶体氨基酸、乙基纤维素包膜赖氨酸、丙烯酸树脂包膜赖氨酸及羧甲基纤维素包膜赖氨酸在不同 pH（6.5、7.0、7.2、7.5、8.0）、不同时间（0.5、1、3 h）、不同温度（20、28 ℃）下的溶失率。结果发现，三种包被材料中，只有羧甲基纤维素在不同时间、不同 pH、不同温度的溶失率基本是最低值，特别是羧甲基纤维素包膜赖氨酸在 pH7.0、20 ℃（水环境）及 pH7.2、28 ℃（鱼体内环境）时的溶失率显著低于其他两种包被材料，表明羧甲基纤维素包膜赖氨酸在水中有效降低了晶体赖氨酸的溶失率。罗运仙等[75]报道晶体赖氨酸经微囊化后，其水中溶失率显著降低，仅相当于晶体赖氨酸的 28.37%。

包膜氨基酸及微囊氨基酸进入水产动物体内后需将包膜材料崩解后才能被吸收，可基本上实现与蛋白态氨基酸的同步吸收。罗运仙等[75]配制了高豆粕组（23%）、低豆粕组（15%）、低豆粕组（15%）+晶体赖氨酸及低豆粕组（15%）+微囊赖氨酸四组饲料，在投喂草鱼 8 周后，对草鱼摄食上述 4 种饲料 0、1、2、3、4、5 h 后的血浆总游离氨基酸浓度测定发现微囊赖氨酸组、高豆粕组、低豆粕组血浆总游离氨基酸含量均在饲后 3 h 达到最大值，而晶体赖氨酸组血浆总游离氨基酸含量出现最大值的时间相对提前 1 h 左右。Zhao 等[76]发现斑点叉尾鲴摄食添加微囊赖氨酸盐酸的饲料后血浆总游离氨基酸浓度在 0、1、2、3、4、5、6 h 的变化趋势与高豆粕组（16%）、低豆粕组（0）相符，在摄食后 4 h 达到峰值，较添加赖氨酸盐酸盐和赖氨酸硫酸盐组延迟 2 h。

包膜或微囊赖氨酸在促进水产动物生长及饲料利用率上的应用效果优于晶体赖氨酸。王冠等[73]在异育银鲫基础饲料中添加硬化油脂包膜赖氨酸、明胶包膜赖氨酸及晶体赖氨酸，发现硬化油脂包膜赖氨酸组、明胶包膜赖氨酸组的增重率较晶体赖氨酸组的增重率分别提高了 18.6%、17.9%，饲料系数下降 0.28、0.37，显著促进了异育银鲫的生长及饲料利用能力。罗运仙等[75]指出在低豆粕饲料分别添加晶体赖氨酸和微囊赖氨酸，发现添加微囊赖氨酸能够使增重率显著提高，使饲料系数显著降低。而谭芳芳等[77]指出草鱼饲料中添加微囊氨基酸可促进鱼体生长、降低饲料系数、增加肌肉粗蛋白含量，但微囊赖酸添加水平超过 0.3%（占饲料）时则会抑制草鱼生长。对罗非鱼的研究表明，饲料中添加包膜 L-Lys·HCl 使鱼体的增重率、特定生长率、饲料转化效率和蛋白质效率、蛋白质沉积率、全鱼粗蛋白含量显著升高，使全鱼粗脂肪、肌肉粗脂肪、肝脏粗脂肪含量显著降低，而普通饲用赖氨酸盐酸盐效果不明显[78]。在冷向军等[79]对鲤的研究中发现与低鱼粉组相比，添加晶体氨基酸对鲤生长性能无显著改善，而添加微囊氨基酸则提高增重率 8.67%，降低饲料系数 7.46%，且在生长性能方面微囊氨基酸组与高鱼粉组没有显著差异。朱选等[80]在低盐度 5~6 和水温 21~26 ℃条件下，以每尾初重 0.79 g 左右的南美白对虾为试验对象，分别投喂基础饲料和添加有赖氨酸盐酸盐、赖氨酸硫酸盐及包膜氨基酸的试验组饲料 48 d 后，发现赖氨酸盐酸盐组、赖氨酸硫酸盐组对生长性能及饲料系数均无显著影响，包膜赖氨酸盐酸盐显著提高特定生长率 15%，但对饲料系数无显著影响。目前关于包膜或微囊赖氨酸的研究主要集中于淡水草食性及杂食性鱼类，而在淡水肉食性鱼类和海水鱼类饲料中的应用效果研究相对缺乏，有待今后进一步开展研究。

四、投喂策略

选择适宜的投喂策略能够在一定程度上增强鱼体对于晶体氨基酸的利用能力。冷向军[81]等在异育银鲫基础饲料中添加晶体 L-赖氨酸和 DL-蛋氨酸，投喂频率设定为 2 次/d、3 次/d、4 次/d，发现 2 次/d、3 次/d 的投喂频率并未明显改善生长性能，而 4 次/d 的投喂频率时则显著提高生长性能，晶体赖氨酸的利用能力也显著增强，表明投喂频率的提高使两次投喂产生的氨基酸峰值相互叠加，增强机体对饲料中氨基酸的利用效率。朱传忠等[82]选用初均重为 22.5 g 的鲫鱼种，分别按照 2 次/d、4 次/d 的投喂频率投喂补充 0.3% 包被赖氨酸的饲料，经过 38 d 的养殖后发现，投喂 4 次/d 较投喂 2 次/d 能够显著提升鲫生长性能，但饲料利用效率并未随投喂频率的提高及包被赖氨酸的补充而显著提升，生长的显著改善在一定程度上归因于对饲料采食量的上升。

五、水温

在水环境中，温度是影响水产动物赖氨酸需要量的重要因素。邓思平等[83]研究发现，当水温逐渐升高，南方鲇的生长速率亦会随之提高，当水温达到 28 ℃时，生长速率达到峰值，若温度开始下降时，生长速率也会随之下降。因此，水产动物赖氨酸的需要量会随着水体环境中温度的改变而改变，即水温较高条件下，赖氨酸需要量会有一定程度的提高，而在水温较低时赖氨酸需要量也会相应降低。Viola 等[84]以 95~120 g 的鲤为研究对象，发现在网箱养殖条件下，当水体温度低于 22 ℃时，添加量超过 1.7% 的赖氨酸并不会促进机体生长；而当水体温度达到 26 ℃时，1.7% 赖氨酸组的特定生长率较未添加组提升了 10%~20%。

第四节 赖氨酸与水产动物健康

一、赖氨酸对水产动物消化吸收能力的影响

赖氨酸影响着水产动物消化器官（消化道及消化腺）的发育，这关系到机体能否发挥

正常消化吸收功能。尤其是对于鲤、草鱼等无胃鱼类，肠道是营养物质消化吸收的最主要场所。在幼建鲤上的研究发现饲料中赖氨酸添加量的提高能够显著提高幼建鲤的肝胰脏及肠道的重量、占体指数和蛋白含量[85]。在草鱼的研究发现草鱼的肝胰脏重量、肝体指数、肠重和肠长均随着赖氨酸水平的增加而提高[86]。上述研究结果表明赖氨酸对水产动物消化吸收能力的改善可通过促进肝胰脏与肠道的相对生长及绝对生长来实现。

赖氨酸影响着水产动物消化道中消化酶的活力。赵春蓉[87]的研究发现赖氨酸从0.7%提高到1.9%能显著提高幼建鲤肠道中蛋白酶、脂肪酶、Na^+/K^+-ATPase酶和碱性磷酸酶活性显著提高。Li等[88]对草鱼的研究指出，饲料赖氨酸水平达到1.29%时，肠道胰蛋白酶、糜蛋白酶、脂肪酶、Na^+/K^+-ATPase及碱性磷酸酶的活力达到最高，过低或过高的赖氨酸水平均会抑制相关消化酶的活力。Naz和Turkmen[89]发现赖氨酸可刺激金头鲷胰脏中胰蛋白酶原的释放，说明赖氨酸既可以促进鱼胰腺消化酶的合成，又可以影响鱼类消化酶的分泌和释放。上述研究结果表明提高消化酶的活力是赖氨酸改善水产动物消化吸收能力的重要途径。

二、赖氨酸对水产动物免疫功能的影响

饲料中添加适宜水平的赖氨酸能够改善鱼体的免疫能力及抗氧化能力。鄢华[85]发现幼建鲤在缺乏赖氨酸时会表现出对病原菌的识别、黏附和溶菌能力的减弱，同时肠道中的有害菌群（如大肠杆菌、嗜水气单胞菌）会随着赖氨酸的缺乏显著增加，而有益菌群（如乳酸杆菌）数量则会显著降低，从而使疾病抵抗能力降低，不利于幼建鲤的生长。赵春蓉[87]对幼建鲤的研究中发现，饲料中添加赖氨酸可增强鱼体免疫能力，即在嗜水气单胞菌攻毒17 d后，赖氨酸组的第17天成活率、攻毒后第17天血清中溶菌酶活力、总抗体水平和抗嗜水气单胞菌抗体效价均显著提高，当赖氨酸水平为1.60%和1.45%时，幼建鲤非特异性和特异性免疫力分别达到最强。Li等[88]发现在草鱼饲料中添加赖氨酸显著提高了谷胱甘肽（glutathione，GSH）的含量，提示赖氨酸能够增强鱼体的非酶抗氧化防御能力。对虹鳟的研究表明，赖氨酸通过谷胱甘肽介导的ε-聚赖氨酸途径进行分解；为维持细胞内还原型谷胱甘肽与氧化型谷胱甘肽的比例相对平衡，赖氨酸可以提升还原型谷胱甘肽的占比；添加赖氨酸还可以显著提高谷胱甘肽过氧化物酶（glutathione peroxidase，GPx），谷胱甘肽还原酶（glutathione reductase，GR）和过氧化氢酶（catalase，CAT）活力[90]。对生长中期的草鱼，赖氨酸可能通过上调生长中期草鱼肠道核转录因子E2相关因子2（nuclear factor E2-related factor 2，$Nrf2$）和下调胞浆蛋白伴侣分子1b（Kelch-like ECH-associated protein，$Keap1b$），上调超氧化物歧化酶（superoxide dismutase，SOD）、GPx、CAT和谷胱甘肽硫转移酶（Glutathione-S-transferase，GST）基因表达水平，增加SOD、GPx、CAT和GST的活力和GSH含量[86]。上述研究表明赖氨酸可通过改善非酶促和酶促抗氧化防御能力，增强体内自由基的清除能力，减少脂质过氧化和蛋白质氧化带来的损伤，进而保证水产动物消化器官（如肝胰脏和肠道等）结构完整性。

关于赖氨酸影响水产动物免疫功能的机制研究还比较缺乏。Yang等[91]对大口黑鲈的研究发现赖氨酸通过调节生长激素-类胰岛素样生长因子生长轴（growth hormone-insulin-like growth factor axis，$GH-IGF$）基因表达水平，不仅调控鱼类生长发育，而

且能够调控机体细胞免疫和体液免疫的功能、抗氧化能力、促炎性反应和抗炎反应。Toll样受体（Toll）-髓样分化因子 88（myeloiddifferentiationfactor88，MyD88）-细胞核因子 kappa B（nuclear factor kappa-B，NFκB）/细胞核因子 κp65 蛋白（RelA）是调控机体免疫应答的重要信号通路，而赖氨酸缺乏会显著增加大口黑鲈头肾此信号通路中 $MyD88$、$RelA$ 和抗生素抗性基因（antibiotic resistance gene，ARG）的基因表达水平，使机体产生免疫应答反应，在饲料中补充赖氨酸则会改善赖氨酸缺乏的免疫应答反应。饲料中补充赖氨酸能够显著降低大口黑鲈促炎性细胞因子（白细胞介素-1β，interleukin-1β，IL-1β；白细胞介素-8，interleukin-8，IL-8）和抗炎细胞因子（转化生长因子-β1，transforming growth factor-β1，TGF-β1；白细胞介素-10，interleukin-10，IL-10），表明赖氨酸通过上调促炎性细胞因子的基因表达激发抗炎反应。

三、赖氨酸对水产动物氮代谢的影响

饲料中赖氨酸的添加主要通过两种途径改善水产动物氮的沉积，即增加水产动物氮摄入量和提升机体对饲料蛋白的利用率，进而提升水产动物蛋白质的沉积效率，减少氮排出，降低环境污染。赖氨酸可以通过以及影响动物对饲料蛋白质的利用效率影响动物的氮沉积。

水产动物对饲料溶出于水中的赖氨酸较为敏感，刺激味觉器官进而激发食欲，增加摄食量[92]。对遮目鱼的研究结果表明，饲料中赖氨酸的缺乏会抑制鱼体食欲，使鱼体摄食量减少，导致鱼体氮的沉积量降低[93]。对鲤[87]、真鲷[93]、日本比目鱼[94]和虹鳟[95]的研究表明，鱼体摄食量会随着饲料赖氨酸水平的提高而提高，从而提高机体氮沉积。但赖氨酸影响水产动物摄食量的具体机制仍需进一步研究。

在需要量范围内赖氨酸水平的增加在提升水产动物的氮沉积率的同时，也明显降低了氮代谢终产物——氨的排出量。付锦锋[54]研究发现在虹鳟饲料中添加赖氨酸，当以植物蛋白作为基础蛋白质来源，外源赖氨酸的添加量为 0.8% 时（饲料赖氨酸水平为 2.1%），与对照组（42% 蛋白组；以鱼粉为主要蛋白源，饲料赖氨酸水平为 2.0%）相比，排入水中的总氨氮和可溶性磷分别减少了 28% 和 71%。对黄尾鱼的研究表明，饲料中添加 1.05%~1.85% 赖氨酸，氮沉积率显著升高，但达到 3.05% 时则不会显著影响氮沉积率[96]。李雪吟[86]发现赖氨酸显著上调了草鱼肠道雷帕霉素靶蛋白（TOR）基因表达水平，显著下调了后肠 4E 结合蛋白（$4E-BP$）基因表达水平，表明赖氨酸可能通过调控 TOR 和 $4E-BP$ 的基因表达，提高机体蛋白质合成能力及消化吸收能力，促进机体对外源赖氨酸的利用能力，造成用于分解供能的氨基酸减少，进而提高水产动物的蛋白质沉积效率。

四、赖氨酸对脂肪代谢及糖代谢的影响

植物蛋白替代鱼粉带来的氨基酸缺乏会明显降低脂质合成代谢率，提高脂质分解代谢率[97]。目前关于赖氨酸调节脂肪代谢的研究相对较少，仅在大口黑鲈[91]上出现过相关报道，赖氨酸缺乏显著降低肝脏大口黑鲈中脂肪酸和甘油三酯含量，同时下调了脂肪生成的关键酶（氧化物酶体增殖物激活受体-α，peroxisome proliferator-activated receptor-α，

$PPAR-\alpha$；氧化物酶体增殖物激活受体-β，peroxisome proliferator - activated receptor - β，$PPAR-\beta$；乙酰辅酶 A 羧化酶 1，acetyl - CoA carboxylase 1，$ACC1$；甘油二酯酰基转移酶，diacylglycerol acyltransferase 1，$DGAT1$；磷脂酸磷酸水解酶，Lipid phosphate phosphohydrolase，$LPIN1$）的基因表达水平，上调了脂肪分解的关键酶（脂蛋白脂肪酶，lipoprotein lipase，LPL；单酰基甘油脂肪酶，Monoacylglycerol lipase，MAGL），表明饲料中添加晶体赖氨酸能有效改善鱼类脂肪生成和抑制脂肪分解。

Lansard 等[90]研究了 L-亮氨酸、L-蛋氨酸、L-赖氨酸对虹鳟肝细胞中糖代谢相关基因表达的影响，发现 3 种氨基酸与胰岛素的组合均未诱导丙酮酸激酶（pyruvate kinase，PK）mRNA 的表达；赖氨酸对葡萄糖激酶（glucokinase，GK）基因表达无影响，而亮氨酸和蛋氨酸则分别激活和下调了 GK 基因表达。而 Yang 等[91]对大口黑鲈的研究却发现，与赖氨酸缺乏组相比，饲料中赖氨酸的添加显著提升了血清中血糖含量 PK 基因表达，对 GK 和磷酸烯醇式丙酮酸羧激酶（phosphoenolpyruvate carboxykinase，$PEPCK$）的基因无显著影响。上述研究存在的差异表明赖氨酸对鱼类糖代谢的调控能力较为有限，可能与不同鱼种、生长阶段、饲料组成、养殖环境等因素有关。

第五节 小结与展望

目前，关于鱼类、虾蟹类的赖氨酸需要量的研究已经取得较大进展。对大多数鱼类来说，赖氨酸需要量都在 1.34%～3.83%（占饲料）的范围内；对于虾类来说，赖氨酸需要量都在 0.77%～2.63%（占饲料）的范围内；对于蟹类来说，赖氨酸需要量都在 2.34%～2.41%（占饲料）的范围内。同时，关于影响水产动物赖氨酸利用能力的因素（如赖氨酸与精氨酸间的颉颃作用，赖氨酸添加形式等）已开展了大量研究；较为系统地研究了赖氨酸缺乏或过量对水产动物的消化吸收能力、免疫功能和氮代谢的研究。

现有的赖氨酸需要量研究主要集中幼鱼阶段，成鱼阶段的赖氨酸需要量研究相对较少，虾类及蟹类的相关研究也呈现出这样的特点，因此未来应开展大规格鱼类、虾类及蟹类赖氨酸需要量研究；饲料中赖氨酸的添加水平能够影响水产动物的蛋白质利用效率及蛋白质沉积效率，但关于赖氨酸对于蛋白质品质及肌肉品质的调控研究相对较少，因此关于赖氨酸对蛋白质品质及肌肉品质影响也应成为未来的研究方向；赖氨酸缺乏会明显降低脂质合成代谢率，提高脂质分解代谢率，表明赖氨酸在一定程度上能够调控机体脂肪代谢的，其具体调控机制及机理有待进一步研究。

参考文献

[1] 张庆功，王建学，卫育良，等. 红鳍东方鲀幼鱼赖氨酸需要量的研究 [J]. 动物营养学报，2020，32（2）：847-855.

[2] Ebeneezar S, Vijaygoapala P, Srivastava P P, et al. Dietary lysine requirement of juvenile Silver pompano, *Trachinotus blochii* (Lacepede, 1801) [J]. Aquaculture, 2019, 511：734234.

[3] 杜强，林黑着，牛津，等. 卵形鲳鲹幼鱼的赖氨酸需要量 [J]. 动物营养学报，2011，23（10）：

1725-1732.

[4] Xie F J, Ai Q H, Mai K S, et al. Dietary lysine requirement of large yellow croaker (*Pseudosciaena crocea*, Richardson 1846) larvae [J]. Aquaculture Research, 2012, 43: 917-928.

[5] Zhang C, Ai Q, Mai K, et al. Dietary lysine requirement of large yellow croaker, *Pseudosciaena crocea* R. [J]. Aquaculture, 2008, 280: 123-127.

[6] 代伟伟. 半滑舌鳎赖氨酸需求及其蛋白源替代研究 [D]. 青岛：中国海洋大学, 2011.

[7] Luo Z, Liu Y, Mai K, et al. Quantitative L-lysine requirement of juvenile grouper *Epinephelus coioides* [J]. Aquaculture nutrition, 2010, 12 (3): 165-172.

[8] 王学武. 两个生长阶段斜带石斑鱼对赖氨酸和蛋氨酸最适需要量的研究 [D]. 湛江：广东海洋大学, 2014.

[9] Khan M A, Abidi S F. Effect of dietary L-lysine levels on growth, feed conversion, lysine retention efficiency and haematological indices of *Heteropneustes fossilis* (Bloch) fry [J]. Aquaculture Nutrition, 2010, 17: 657-667.

[10] Farhat, Khan M A. Dietary L-lysine requirement of fingerling stinging catfish, *Heteropneustes fossilis* (Bloch) for optimizing growth, feed conversion, protein and lysine deposition [J]. Aquaculture Research, 2013, 44 (4): 523-533.

[11] Yang S D, Liu F G, Liou C H. Assessment of dietary lysine requirement for silver perch (*Bidyanus bidyanus*) juveniles [J]. Aquaculture, 2011, 312: 102-108.

[12] Gridale-Helland B, Hatlen B, Mundheim H, et al. Dietary lysine requirement and efficiency of lysine utilization for growth of Atlantic cod [J]. Aquaculture, 2011, 315 (3-4): 260-268.

[13] Gridale-Helland B, Gatlin Ⅲ D M, Corrent E, et al. The minimum dietary lysine requirement, maintenance requirement and efficiency of lysine utilization for growth of Atlantic salmon smolts [J]. Aquaculture Research, 2011, 42 (10): 1509-1529.

[14] Dominy W, Koshio S, Muashig R. Dietary lysine requirement of juvenile Pacific threadfin (*Polydactylus sexfilis*) [J]. Aquaculture, 2010, 308: 44-48.

[15] Zhou F, Shao J, Xu R, et al. Quantitative L-lysine requirement of juvenile black sea bream (*Sparus macrocephalus*) [J]. Aquaculture Nutrition, 2009, 16 (2): 194-204.

[16] Zhou Q C, Wu Z H, Chi S Y, et al. Dietary lysine requirement of juvenile cobia (*Rachycentron canadum*) [J]. Aquaculture, 2007, 273: 634-640.

[17] 周萌, 吴建开, 曹俊明, 等. 军曹鱼幼鱼对赖氨酸需要量的初步研究 [J]. 长江大学学报（自科版）农学卷, 2005, (1): 50-52, 10.

[18] Mai K S, Zhang L, Ai Q H, et al. Dietary lysine requirement of juvenile Japanese seabass, *Lateolabrax japonicus* [J]. Aquaculture, 2006, 258: 535-542.

[19] 门珂珂. 饲料玉米蛋白粉及赖氨酸对鲈鱼生长和代谢的影响 [D]. 青岛：中国海洋大学, 2014.

[20] 严全根, 解绶启, 雷武, 等. 许氏平鲉幼鱼的赖氨酸需要量 [J]. 水生生物学报, 2006, (4): 459-465.

[21] Marcouli P A, Alexis M N, Andriopoulou A, et al. Dietary lysine requirement of juvenile gilthead seabream *Sparus aurata* L. [J]. Aquaculture Nutrition, 2006, 12 (1): 25-33.

[22] Murillo-Gurrea D P, Coloso R M, Borlongan I G, et al. Lysine and arginine requirements of juvenile Asian sea bass (*Lates calcarifer*) [J]. Journal of Applied Ichthyology, 2001, 17 (2): 49-53.

[23] Small B C, Jr Soares J H. Quantitative dietary lysine requirement of juvenile striped bass *Morone saxatilis* [J]. Aquaculture Nutrition, 2000, 6: 207-212.

[24] Lee S, Small B C, Patro B, et al. The dietary lysine requirement for optimum protein retention differs with rainbow trout (*Oncorhynchus mykiss* Walbaum) strain [J]. Aquaculture, 2019, 511: 7344834.

[25] Prabu E, Felix N, Uma A, et al. Effects of dietary L‐lysine supplementation on growth, body composition and muscle‐growth‐related gene expression with an estimation of lysine requirement of GIFT tilapia [J]. Aquaculture Nutrition, 2019, 26: 568‐578.

[26] 尹东鹏, 陈秀梅, 刘丹妮, 等. 乌鳢饲料赖氨酸及其他必需氨基酸营养需要量的研究 [J]. 饲料工业, 2018, 39 (14): 18‐23.

[27] Adesola A A, Jones C L W, Shipton T A. Dietary lysine requirement of juvenile dusky kob, *Argyrosomus japonicus* [J]. Aquaculture Nutrition, 2018, 24: 673‐680.

[28] 左亚男, 郑伟, 夏长革, 等. 饲料赖氨酸水平对雅罗鱼生长、饲料利用及肌肉营养成分的影响 [J]. 饲料工业, 2017, 38 (8): 29‐35.

[29] Michelato M, De Oliveira Vidal L V, Xavier T O, et al. Dietary lysine requirement to enhance muscle development and fillet yield of finishing Nile tilapia [J]. Aquaculture, 2016, 2457: 124‐130.

[30] 廖英杰, 刘波, 任鸣春, 等. 赖氨酸对团头鲂幼鱼生长、血清生化及游离必需氨基酸的影响 [J]. 水产学报, 2013, 37 (11): 1716‐1724.

[31] 宋长友, 任鸣春, 谢骏, 等. 不同生长阶段团头鲂的赖氨酸需要量研究 [J]. 上海海洋大学学报, 2016, 25 (3): 396‐405.

[32] 王裕玉. 乌苏里拟鲿 *Pseudobagras ussuriensis* 品质评价、适宜蛋白能量水平和氨基酸需要量及对豆粕的利用研究 [D]. 哈尔滨: 东北农业大学, 2013.

[33] 杨威. 鳜幼鱼赖氨酸需要量的研究 [D]. 武汉: 华中农业大学, 2012.

[34] Lin Y, Gong Y, Yuan Y, et al. Dietary L‐lysine requirement of juvenile Chinese sucker, *Myxocyprinus asiaticus* [J]. Aquaculture Research, 2013, 44 (10): 1539‐1549.

[35] Cao J M, Chen Y, Zhu X, et al. A study on dietary L‐lysine requirement of juvenile yellow catfish *Pelteobagrus fulvidracol* [J]. Aquaculture Nutrition, 2012, 18 (1): 35‐45.

[36] 邱红, 黄文文, 侯迎梅, 等. 黄颡鱼幼鱼的赖氨酸需要量 [J]. 动物营养学报, 2015, 27 (10): 3057‐3066.

[37] Abimorad E G, Favero G C, Squassoni G H, et al. Dietary digestible lysine requirement and essential amino acid to lysine ratio for pacu *Piaractus mesopotamicus* [J]. Aquaculture Nutrition, 2010, 16 (4): 370‐377.

[38] Bicudo Á J A, Sado R Y, Cyrino J E P. Dietary lysine requirement of juvenile pacu *Piaractus mesopotamicus* (Holmberg, 1887) [J]. Aquaculture, 2009, 297: 151‐156.

[39] Zhou X Q, Zhao C R, Jiang J, et al. Dietary lysine requirement of juvenile Jian carp (*Cyprinus carpio* var. *jian*) [J]. Aquaculture nutrition, 2008, 14 (5): 381‐386.

[40] 周贤君, 解绶启, 谢从新, 等. 异育银鲫幼鱼对饲料中赖氨酸的利用及需要量研究 [J]. 水生生物学报, 2006, (3), 247‐255.

[41] 王鑫, 薛敏, 王嘉, 等. 养成期异育银鲫对饲料赖氨酸的需要量 [J]. 动物营养学报, 2014, 26 (7): 1864‐1872.

[42] Wang S, Liu Y J, Tian L X, et al. Quantitative dietary lysine requirement of juvenile grass carp *Ctenopharyngodon idella* [J]. Aquaculture, 2015, 249: 419‐429.

[43] Ahmed I, Khan M A. Dietary lysine requirement of fingerling Indian major carp, *Cirrhinus mrigala* (Hamilton) [J]. Aquaculture, 2014, 235: 499‐511.

[44] Tantikitti C, Chimsung N. Dietary lysine requirement of freshwater catfish (*Mystus nemurus* Cuv. &

Val.) [J]. Aquaculture Research, 2002, 32 (s1): 135-141.

[45] Palma J, Andrade J P, Lemme A, et al. Quantitative dietary requirement of juvenile Atlantic ditch shrimp, *Palaemonetes varians*, for lysine, methionine and arginine [J]. Aquaculture Research, 2015, 46 (8): 1822-1830.

[46] 孙丽慧, 沈斌乾, 陈建明, 等. 日本沼虾对饲料赖氨酸的需要量研究 [J]. 上海海洋大学学报, 2013, 22 (1): 100-104.

[47] Xie F, Zeng W, Zhou Q, et al. Dietary lysine requirement of juvenile Pacific white shrimp, *Litopenaeus vannamei* [J]. Aquaculture, 2012, 358-359: 116-121.

[48] 曾雯娉. 凡纳滨对虾幼虾对赖氨酸、蛋氨酸、精氨酸和苯丙氨酸需要量的研究 [D]. 湛江: 广东海洋大学, 2012.

[49] 张微微. 克氏原螯虾对赖氨酸和精氨酸需要量的研究 [D]. 南京: 南京农业大学, 2012.

[50] Richard L, Blanc P P, Rigolet V, et al. Maintenance and growth requirements for nitrogen, lysine and methionine and their utilisation efficiencies in juvenile black tiger shrimp, *Penaeus monodon*, using a factorial approach [J]. British Journal of Nutrition, 2010, 103 (7): 984.

[51] 金敏. 三疣梭子蟹幼蟹对蛋白质、精氨酸、赖氨酸和蛋氨酸需要量的研究 [D]. 宁波: 宁波大学, 2014.

[52] 叶金云, 王友慧, 郭建林, 等. 中华绒螯蟹对赖氨酸、蛋氨酸和精氨酸的需要量 [J]. 水产学报, 2010, 34 (10): 1541-1548.

[53] Fournier V, Gouillou-Coustans M, Metailler R, et al. Excess dietary arginine affects urea excretion but does not improve N utilisation in rainbow trout *Oncorhynchus mykiss* and turbot *Psetta maxima* [J]. Aquaculture, 2003, 217: 559-576.

[54] 付锦锋. 赖氨酸和蛋白水平对虹鳟生长、氨氮和磷排放的影响 [J]. 饲料研究, 2012 (2): 4-6, 10.

[55] Thu T T N, Parkouda C, Saeger S D, et al. Comparison of the lysine utilization efficiency in different plant protein sources supplemented with L-lysine·HCl in rainbow trout (*Oncorhynchus mykiss*) fry [J]. Aquaculture, 2007, 272: 477-488.

[56] Carew L B, Evarts K G, Alster F A. Growth, feed intake, and plasma thyroid hormone levels in chicks fed dietary excesses of essential amino acids [J]. Poultry Science, 1998, 77: 295-298.

[57] Jones J D, Wolters R, Burnett P C. Lysine-arginine-electrolyte relationships in the rat [J]. The Journal of Nutrition, 1966, 89: 171.

[58] Robinson E H, Wilson R P, Poe W E, et al. Arginine requirement and apparent absence of a lysine-arginine antagonist in fingerling channel catfish [J]. The Journal of Nutrition, 1981, 111 (1): 46-52.

[59] Kim K I, Kaye T B, Amundson C H. Requirements for lysine and arginine by rainbow trout (*Oncorhynchus mykiss*) [J]. Aquaculture, 1992, 106 (3-4): 333-344.

[60] Twibell R G, Brown P B. Dietary arginine requirement of juvenile yellow perch [J]. The Journal of Nutrition, 1997, 127 (9): 1838-1841.

[61] 代伟伟, 麦康森, 徐玮, 等. 饲料中赖氨酸和精氨酸含量对大菱鲆幼鱼生长、体成分和肌肉氨基酸含量的影响 [J]. 水产学报, 2015, 39 (6): 876-887.

[62] Palma J, Andrade J P, Lemme A, et al. Quantitative dietary requirement of juvenile Atlantic ditch shrimp *Palaemonetes varians* for lysine, methionine and arginine [J]. Aquaculture Research, 2015, 46: 1822-1830.

[63] Dong X J, Wu J, Shen Y, et al. Effects of different arginine/lysine level on growth performance,

[64] Friedman M, Gumbmmann M R. Bioavailability of some lysine derivatives in mice [J]. Journal of Nutrition, 1981, 111: 1362-1369.

[65] Deng J, Zhang X, Tao L, et al. D-lysine can be effectively utilized for growth by common carp (*Cyprinus carpio*) [J]. Aquaculture Nutrition, 2010, 17 (2): 467-475.

[66] 赵金鑫, 李小勤, 彭松, 等. 斑点叉尾鮰对不同形式赖氨酸利用的比较研究 [J]. 水生生物学报, 2016, 40 (1): 19-26.

[67] Rodehutscord M, Borchert F, Gregus Z, et al. Availability and utilisation of free lysine in rain-bow trout (*Oncorhynchus mykiss*): 2. Comparison of L-lysine·HCl and L-lysine sulphate [J]. Aquaculture, 2000, 187 (1/2): 177-183.

[68] 胡凯, 苏玥宁, 冯琳, 等. 80%赖氨酸硫酸盐与98%赖氨酸盐酸盐对生长中期草鱼生长性能、消化吸收能力和消化器官生长发育影响的比较研究 [J]. 动物营养学报, 2017, 29 (12): 4372-4385.

[69] Niu J, Chen X, Lin H Z, et al. Comparison of L-lysine·HCl and L-lysine sulphate in the feed of *Penaeus monodon* and reevaluation of dietary lysine requirement for *P. monodon* [J]. Aquaculture Research, 201748 (1): 134-148.

[70] 赵金鑫. 斑点叉尾鮰对不同形式赖氨酸和蛋氨酸利用的比较研究 [D]. 上海: 上海海洋大学, 2015.

[71] 刘永坚, 田丽霞, 刘栋辉, 等. 实用饲料补充结晶或包膜赖氨酸对草鱼生长、血清游离氨基酸和肌肉蛋白质合成率的影响 [J]. 水产学报, 2002, 26 (3): 252-258.

[72] Alam M S, Tesshimas I, Koshio S, et al. Supplemental effects of coated methionine and/or lysine to soy protein isolate diet for juvenile kuruma shrimp, *Marsupenaeus japonicus* [J]. Aquaculture, 2005, 248 (1): 13-19.

[73] 王冠, 冷向军, 李小勤, 等. 饲料中添加包膜氨基酸对异育银鲫生长和体成分的影响 [J]. 上海水产大学学报, 2006, 15 (3): 365-369.

[74] 钱前. 晶体或包膜赖氨酸对湘云鲫生长及生理效应的影响 [D]. 重庆: 西南大学, 2014.

[75] 罗运仙, 谢骏, 吕利群, 等. 饲料中补充晶体或微囊赖氨酸对草鱼生长和血浆总游离氨基酸的影响 [J]. 水产学报, 2010, 4 (3): 466-473.

[76] Zhao J X, Li X Q, Leng X J, et al. Comparative study on the utilization of different lysine sources by channel catfish (*Ictalurus punctatus*) [J]. Aquaculture Nutrition, 2017, 23: 833-843.

[77] 谭芳芳, 叶元土, 肖顺应, 等. 补充微囊赖氨酸和蛋氨酸对草鱼生长性能的影响 [J]. 动物营养学报, 2010, 22 (3): 804-810.

[78] 郑严严. 包膜赖氨酸盐酸盐的制备及其对罗非鱼生长性能、饲料利用的影响和作用机理的探讨 [D]. 杭州: 浙江大学, 2009.

[79] 冷向军, 罗运仙, 李小勤, 等. 饲料中添加晶体或微囊氨基酸对鲤生长性能的影响 [J]. 动物营养学报, 2010, 22 (6): 1599-1606.

[80] 朱选, 曹俊明, 蓝汉冰, 等. 饲料中添加不同形式赖氨酸对南美白对虾生长及肌肉成分影响的研究 [J]. 饲料工业, 2008, (4): 16-18.

[81] 冷向军, 王冠. 投饲频率对异育银鲫饲料中添加晶体氨基酸的影响 [J]. 饲料研究, 2006, (12): 50-52.

[82] 朱传忠, 杨健, 骆作勇. 不同投喂频率下补充包被氨基酸对鲫鱼生长性能的影响 [J]. 饲料工业, 2011, (S2): 54-57.

[83] 邓思平,吴天利,王德寿,等. 温度对南方鲇幼鱼生长与发育的影响 [J]. 西南师范大学学报(自然科学版),2000,(6):674-679.

[84] Viola S, Lahav E, Arieli Y. Response of Israeli carp, *Cyprinus carpio* L. to lysine supplementation of a practical ration at varying conditions of fish size, temperature, density and ration size [J]. Aquaculture Research, 1992, 23 (1): 49-58.

[85] 鄢华. 赖氨酸缺乏对幼建鲤肠道菌群、消化酶活力和免疫功能的影响 [D]. 成都:四川农业大学, 2007.

[86] 李雪吟. 赖氨酸对草鱼肠道免疫和结构屏障作用及其机制研究 [D]. 成都:四川农业大学, 2017.

[87] 赵春蓉. 赖氨酸对幼建鲤消化能力和免疫功能的影响 [D]. 成都:四川农业大学, 2005.

[88] Li X Y, Tang L, Hu K, et al. Effect of dietary lysine on growth, intestinal enzymes activities and antioxidant status of sub-adult grass carp (*Ctenopharyngodon idella*) [J]. Fish Physiology and Biochemistry, 2013, 40 (3): 659-671.

[89] Naz M, Turkmen M. Changes in the digestive enzymes and hormones of gilthead seabream larvae (*Sparus aurata*, L.1758) fed on *Artemia nauplii* enriched with free lysine [J]. Aquaculture International, 2008, 17 (6): 523-535.

[90] Lansard M, Panserat S, Plangnes-Juan E, et al. L-Leucine, L-Methionine, and L-Lysine Are involved in the regulation of intermediary metabolism-related gene expression in rainbow trout hepatocytes [J]. Journal of Nutrition, 2011, 141 (1): 75-80.

[91] Yang P, Wang W Q, Chi S Y, et al. Effects of dietary lysine on regulating GH-IGF system, intermediate metabolism and immune response in largemouth bass (*Micropterus salmoides*) [J]. Aquaculture Reports, 2020, 17: 100323.

[92] 曾端,杨春贵. 水产动物的摄食化学感受器及水产诱食剂的开发应用 [J]. 水产养殖, 2002, (3): 15-17.

[93] Borlongan I, Benitez L V. Quantitative lysine requirement of milkfish juveniles [J]. Aquaculture, 1990, 87: 341-347.

[94] Forster I, Yogata H. Lysine requirement of juvenile Japanese flounder *Paralichthys olioaceus* and juvenile red sea bream *Pagrus major* [J]. Aquaculture, 1998, 161: 131-142.

[95] Encamac P, De Lange C, Rodehutsord M, et al. Diet digestible energy content affects lysine utilization, but not dietary lysine requirements of rainbow trout (*Oncorhynchus mykiss*) for maximum growth [J]. Aquaculture, 2004, 235: 569-586.

[96] Ruchmat T, Masumoto T, Itoh H Y, et al. Quantitative lysine requirement of yellowtail [J]. Aquaculture, 1997, 158: 331-339.

[97] Song F, Xu D D, Mai K S, et al. Comparative study on the cellular and systemic nutrient sensing and intermediary metabolism after partial replacement of fishmeal by meat and bone meal in the diet of turbot (*Scophthalmus maximus* L.) [J]. PloS One, 2016, 11: 1-19.

第四章 蛋氨酸

鱼粉是水产动物的优质蛋白饲料原料，但随着水产养殖业的发展，鱼粉的供给量已不能满足饲料工业的需求，因此，使用植物蛋白原料高比例替代饲料中鱼粉已经成为近年来水产动物营养研究的热点。但由于植物蛋白原料的氨基酸组成与鱼粉间存在较大差异，高比例替代会导致饲料氨基酸不平衡，进而降低水产动物的生长性能、饲料利用率及免疫力。蛋氨酸是高比例替代鱼粉的第一限制性氨基酸，添加蛋氨酸及其类似物可改善饲料氨基酸平衡，提高饲料利用率，促进生长，改善健康。本章综述了蛋氨酸代谢途径、水产动物蛋氨酸的需要量、蛋氨酸与其他营养素的相互作用及对水产动物健康的影响，以期为蛋氨酸在水产动物中开展深入研究提供参考。

第一节 蛋氨酸的理化性质

蛋氨酸（Methionine，Met），分子式 $C_5H_{11}O_2NS$，分子质量 150.22，分子结构如图 4-1 所示，熔点为 280～281 ℃，外观为白色薄片状结晶或结晶性粉末，溶于水、稀酸和稀碱，极微溶于 95% 的乙醇，极难溶于无水乙醇，几乎不溶于乙醚。有特殊气味，味微甜。

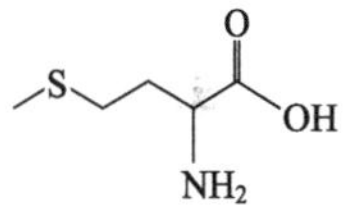

图 4-1 蛋氨酸分子结构

第二节 蛋氨酸的代谢途径

氨基酸通过肠道吸收后进入肝静脉，转运至肝脏进行代谢。蛋氨酸在肝脏中可以合成半胱氨酸、胆碱和牛磺酸，同时也是机体细胞代谢的甲基供体。蛋氨酸是 S-腺苷甲硫氨酸（S-adenosyl-L-methionine，SAM）的前体物质，SAM 转化生成 S-腺苷同型半胱氨酸（S-adenosylhomocysteine，SAH），但由于 SAH 不稳定进一步转换为同型半胱氨酸。同型半胱氨酸具有细胞毒性，需进一步通过蛋氨酸合成酶（methionine synthase，MS）或半胱氨酸甲基转移酶（betaine-homocysteine methyltransferase，BHMT）进行再次甲基化，或通过转硫作用途径清除[1]。蛋氨酸和三磷酸腺苷（ATP）调控 SAM 的合成，SAM 作为甲基供体参与机体胆碱的生物合成。饲料充足的蛋氨酸可以促进肝脏胆碱和肉碱的合成，并为胆固醇和脂质蛋白的合成提供充足的磷脂质和乙酰辅酶 A[2]。此外，SAM 还可以通过脱羧基作用生成正丙基，参与多胺合成。多胺对细胞生存和生长具有重要作用[3]。蛋氨酸分解代谢生成的同型半胱氨酸，在胱硫醚合成酶（cystathiorinesynthase）、胱硫醚酶（cystathionine）催化下生成 α-酮丁酸，α-酮丁酸进一步转换为琥珀酰 CoA。琥珀酰 CoA 经三羧酸循环转变成草酰乙酸，然后经磷酸烯醇式丙酮酸羧激酶（PEPCK）催化生成磷酸烯醇式丙酮酸，后者经丙酮酸转变为乙酰 CoA，其进入三羧酸循环被彻底氧化为 CO_2 和 H_2O。

第三节 水产动物对蛋氨酸的需要量

蛋氨酸是水产动物正常生长、发育的必需氨基酸，在蛋白质合成、生成半胱氨酸中发挥重要作用。例如，虹鳟蛋氨酸缺乏组蛋白质沉积率仅为正常及过量组的 50%，显著降低机体干物质和脂肪含量[4]。水产动物对蛋氨酸的需要量见表 4-1，鱼类对蛋氨酸需要量为 0.73%～1.71%，虾蟹类对蛋氨酸的需要量为 0.70%～1.28%。

表 4-1 水产动物蛋氨酸需要量

	品种	体重（g）	模型	指标	需要量（占饲料干物质,%）	参考文献
鱼类	军曹鱼	9.79	二次多项式模型	增重率	1.24	Chi 等[11]
	杂交石斑鱼	10.61	渐近回归模型	增重率	1.45	Li 等[12]
	斜带石斑鱼	9.75	二次多项式模型	特定生长率	1.40	杨炟懿等[13]
		102.60	二次多项式模型	特定生长率	1.07	
	松浦镜鲤	24.42	折线模型	特定生长率	0.90	程龙等[14]
	黄河鲤	21.56	二次多项式模型	增重率	1.22	Yun 等[15]
				肌肉硬度	0.95	
				肌肉咀嚼性	1.10	
				肌肉弹性	1.37	
	红鳍东方鲀	13.83	二次多项式模型	特定生长率	1.38	张庆功等[16]
	洛氏鱥	13.63	折线模型	特定生长率	1.43	段晶等[17]
	吉富罗非鱼	8.95	二次多项式模型	增重率	0.99	He 等[9]
	大黄鱼	2.00	二次多项式模型	增重率	1.15	Elmada 等[18]
	吉富罗非鱼	2.30	二次多项式模型	增重率	0.91	He 等[19]
	异育银鲫	6.70	二次多项式模型	特定生长率	0.71	Ren 等[20]
	异育银鲫	51.00	二次多项式模型	增重率	0.98	Wang 等[10]
			折线模型	增重率	0.73	
	军曹鱼	150.9	二次多项式模型	增重率	1.04	Wang 等[21]
				饲料转化率	1.15	
	乌苏拟鲿	0.60	二次多项式模型	增重率	1.41	Wang 等[22]
	胭脂鱼	1.72	二次多项式模型	特定生长率	1.41	Chu 等[23]
	团头鲂	3.34	折线模型	特定生长率	0.85	Liao 等[24]
				蛋白质沉积率	0.84	
	印度鲇	4.33	二次多项式模型	增重率	1.49	Ahmed[25]

(续)

	品种	体重（g）	模型	指标	需要量（占饲料干物质,%）	参考文献
鱼类	非洲鲇	78.00	折线模型	日增重	1.90	Elesho 等[26]
			二次多项式模型	日增重	2.92	
			折线模型	饲料系数	1.88	
			二次多项式模型	饲料系数	2.97	
	吉富罗非鱼	66.76	二次多项式模型	增重率	1.13	向枭等[27]
				蛋白质沉积率	1.16	
	卵形鲳鲹	12.40	折线模型	增重率	1.06	Niu 等[28]
				氮沉积率	1.27	
	大菱鲆	5.60	二次多项式模型	特定生长率	1.58	Ma 等[29]
	异育银鲫	2.60	折线模型	蛋氨酸沉积率	0.89	贾鹏等[30]
	草鱼	259.00	二次多项式模型	特定生长率	1.04	唐炳荣等[31]
	草鱼	0.36	折线模型	特定生长率	1.03	Ji 等[32]
	黑鲷	14.21	二次多项式模型	特定生长率	1.71	Zhou 等[33]
	欧洲鲈	13.40	折线模型	氮沉积率	0.80	Tulli 等[34]
				增重率	0.91	
	军曹鱼	11.61	二次多项式模型	特定生长率	1.19	Zhou 等[35]
	斜带石斑鱼	13.25	折线模型	增重率	1.31	Luo 等[36]
虾蟹类	凡纳滨对虾	1.98	二次多项式模型	增重率	0.94	Facanha 等[5]
	凡纳滨对虾	0.98	二次多项式模型	增重率	0.87	Wang 等[37]
	日本沼虾	0.30	二次多项式模型	增重率	0.70	朱杰等[38]
	克氏原螯虾	9.80	二次多项式模型	增重率	0.94	朱杰等[39]
	中华鳖	17.63	二次多项式模型	特定生长率	1.28	周小秋等[40]
	中华绒螯蟹	2.03	二次多项式模型	净增重	1.12	叶金云等[41]
	三疣梭子蟹	11.27	折线模型	特定生长率	1.07	金敏等[42]

水产动物蛋氨酸需要量的不同主要由不同养殖密度、养殖方式、养殖温度、饲料营养水平、统计方法等因素造成。养殖密度影响凡纳滨对虾蛋氨酸的需要量，在养殖密度为 50 和 100 尾/m² 时，蛋氨酸需要量为 0.72%；在养殖密度为 75 尾/m² 时，蛋氨酸需要量为 0.81%[5]。养殖方式也影响凡纳滨对虾蛋氨酸的需要量，在静水养殖和流水养殖条件下，1.98 g 凡纳滨对虾蛋氨酸需要量分别为 0.94%、0.80%[6]。此外，凡纳滨对虾投喂频率由 4 次/d 增加至 6 次/d 时，其增重率、体蛋白的沉积和总蛋白酶的活性得到改善，增加投喂频率有利于提高游离氨基酸的利用效率，进而影响蛋氨酸的需要量[7]。军曹鱼在养殖温度为 30、34 ℃条件下，分别投喂添加 0.05%、0.55%、0.95%蛋氨酸的饲料，研究结果表明，在不同温度条件下添加 0.55%蛋氨酸生长性能均高于 0.05%和 0.95%组，相同蛋氨酸水平条件下，30 ℃生长性能优于 34 ℃，且末重蛋氨酸和温度存在交互作用[8]。饲料

能量水平影响罗非鱼的蛋氨酸需要量，分别配制消化能为 10.9 MJ/kg、12.4 MJ/kg 的低消化能和高消化能饲料，蛋氨酸的需要量分别为 0.73%、0.99%，上述研究表明高消化能组蛋氨酸的需要量高于低消化能组，且在高消化能组增重率高于低消化能组[9]。虹鳟分别投喂蛋氨酸缺乏、正常及过量饲料 12 周，投喂蛋氨酸缺乏饲料 3、6 周后，增重率、饲料转化率均显著低于正常组及过量组，但正常组和过量组差异没有达到显著水平；投喂蛋氨酸过量饲料 9、12 周后，增重率、饲料转化率显著高于缺乏组，但增重率显著低于对照组，表明投喂含蛋氨酸饲料的时间长短影响蛋氨酸的需要量[4]。不同统计方法也会对需要量产生影响，在异育银鲫研究中表明，以增重率为指标，通过三次多项式模型、折线模型得出 51 g 异育银鲫蛋氨酸的需要量分别为 0.98%、0.73%[10]。

第四节　蛋氨酸与其他营养素的相互作用

一、蛋氨酸与牛磺酸之间的相互作用

牛磺酸已经被证明是水产动物的条件性必需营养素，是植物蛋白高比例替代鱼粉饲料限制生长的主要因素。饲料蛋氨酸含量影响水产动物牛磺酸的需要量。黄尾鰤研究结果表明，添加 0.2% 或 1% 的蛋氨酸时，以特定生长率为指标，牛磺酸需要量分别为 7.7%、1.09%；双因素方差分析结果表明，牛磺酸或蛋氨酸均提高增重率、特定生长率，降低饲料系数，且 1% 蛋氨酸组增重率均高于 0.2% 蛋氨酸组[43]。在牛磺酸含量为 0.36%，含硫氨基酸含量为 1.15% 的虹鳟无鱼粉饲料中分别添加 0、0.05%、1% 牛磺酸和 0、0.05%、1% 蛋氨酸，研究结果表明，添加牛磺酸可以显著促进生长、提高采食量；添加蛋氨酸反而降低生长性能和采食量，降低半胱氨酸双加氧酶（cysteine dioxygenase，CDO）和半胱亚磺酸脱羧酶（cysteine sulfinate decarboxylase，CSAD）的基因表达量，生长性能的降低可能与采食量、血清胰岛素样生长因子-Ⅰ（Insulin-like growth factor Ⅰ，IGF-Ⅰ）的降低相关[44]。罗非鱼蛋氨酸缺乏饲料中分别添加 0.4% 的牛磺酸、0.4% 的蛋氨酸，研究结果表明，添加牛磺酸和蛋氨酸都显著提高生长性能和 S-腺苷高半胱氨酸酶的基因表达量，但均显著低于鱼粉组[45]。在大西洋白姑鱼的研究中得到相似的结果，在蛋氨酸缺乏饲料中添加 1% 牛磺酸显著提高粗脂肪消化率和生长性能，提高血清谷草转氨酶、谷丙转氨酶的活性，提高血清胆固醇、总蛋白、甘油三酯及胆汁酸含量；添加 0.26% 的蛋氨酸提高干物质、粗蛋白、粗脂肪及能量的消化率，但对生长性能的影响没有达到显著水平；添加牛磺酸或蛋氨酸均提高果糖-1，6-二磷酸酶（Fructose-1，6-diphosphatase，FBPase）的活性[46,47]。此外，也有研究表明，在蛋氨酸缺乏饲料中添加牛磺酸对欧洲鲈、尖吻鲈的生长、饲料利用率及机体组成没有影响，但在蛋氨酸缺乏饲料中添加牛磺酸显著提高肝脏过氧化氢酶活性[48,49]。

二、蛋氨酸与赖氨酸之间的相互作用

低鱼粉饲料中添加蛋氨酸和赖氨酸具有促进生长的作用。在草鱼豆粕替代鱼粉饲料中，同时添加蛋氨酸和赖氨酸可显著提高增重率，降低饲料系数，降低前肠、中肠、后肠褶皱高度，降低脂肪酶、蛋白酶、淀粉酶、γ-谷氨酰转肽酶、碱性磷酸酶、肌酸激酶的

活性[50]。在大豆浓缩蛋白、肉骨粉、味精蛋白替代21.2%鱼粉的黑鲷饲料中同时添加0.5%赖氨酸和0.34%蛋氨酸，显著提高肝脏、中肠蛋白酶活性和前肠脂肪酶活性，进而提高饲料消化率[51]。对卵形鲳鲹的研究表明，在30%豆粕替代鱼粉饲料中同时添加蛋氨酸和赖氨酸促进生长，提高成活率[52]。在豆粕替代60%鱼粉的花鲈饲料中同时添加0.2%赖氨酸和0.3%蛋氨酸对增重率的影响没有达到显著水平，但饲料系数显著升高；在替代80%的鱼粉组中分别添加0.2%赖氨酸+0.3%蛋氨酸及0.2%赖氨酸+0.7%蛋氨酸增重率均显著降低[53]。上述研究表明，在高比例鱼粉替代饲料中同时添加赖氨酸和蛋氨酸具有提高肠道消化酶活性、提高饲料利用率及促进生长的作用。因此，在用植物蛋白或其他动物蛋白替代鱼粉时，除了要考虑补充蛋氨酸外，还要考虑其他氨基酸的平衡。

三、蛋氨酸与其他营养素的相互作用

胆碱和蛋氨酸均为机体的甲基供体，蛋氨酸缺乏时，胆碱具有一定补偿作用。鱼类蛋氨酸-胆碱缺乏增加肝脏甘油三酯的含量。对大西洋鲑的研究表明，在蛋氨酸缺乏饲料中添加胆碱后肝脏的甘油三酯并没有增加，但肝脏和肌肉总磷脂的含量显著提高[54,55]。Michael等[56]在以大豆分离蛋白为主要蛋白源的蛋氨酸缺乏饲料中，分别添加0、0.06%、0.12%氯化胆碱和0%、1.5%蛋氨酸，研究结果表明，同时添加氯化胆碱和蛋氨酸可以提高日本对虾成活率、增重率及饲料转化率，且在蛋氨酸缺乏饲料的中添加0.12%氯化胆碱能够补偿机体对甲基的需求。

露斯塔野鲮蛋氨酸需要量为1.2%，在蛋氨酸需要量1.5倍的饲料添加2%的岩藻多糖显著提高增重率、特定生长率及饲料转化率，通过提高溶菌酶活性、免疫球蛋白、吞噬活性等，进而提高免疫功能，提高嗜水气单胞菌攻毒后成活率[57]。在豆粕替代60%鱼粉的胭脂鱼饲料中添加0.3%蛋氨酸和1 500 U/kg植酸酶，能够显著提高增重率和磷的表观消化率[58]。对大西洋鲑的研究表明，在植物蛋白源饲料中同时添加0.312%蛋氨酸、0.017 9%维生素B_{12}、0.005 3%叶酸、0.010 7%维生素B_6可提高生长性能及肥满度[59]。

第五节 不同类型蛋氨酸的生物效价

一、包膜蛋氨酸的生物效价

不同类型蛋氨酸的生物效价及对水产动物生长、免疫的作用效果不同。为减少蛋氨酸在水中的溶失率，Alam等[60,61]研究了羧甲基纤维素、玉米蛋白、κ-卡拉胶、琼脂、羧甲基纤维素-玉米蛋白、酪蛋白-明胶、玉米蛋白-κ-卡拉胶、酪蛋白-明胶-κ-卡拉胶包被蛋氨酸饲料中蛋氨酸在硼酸盐缓冲剂中的溶失率，除含玉米蛋白包被蛋氨酸饲料中蛋氨酸溶失率与未包被蛋氨酸饲料差异不显著外，其他组饲料蛋氨酸溶失率均显著降低，浸入硼酸盐缓冲剂中10 min和30 min时，蛋氨酸溶失率最低的为酪蛋白-明胶-κ-卡拉胶和酪蛋白-明胶包被的蛋氨酸饲料，不同包被蛋氨酸饲料蛋氨酸溶失率见表4-2[60]。日本对虾研究结果表明，包被蛋氨酸生长性能优于未包被蛋氨酸，且酪蛋白-明胶包被蛋氨酸生长性能最好。军曹鱼低鱼粉饲料中添加包膜蛋氨酸有助于蛋氨酸在肠道中的缓慢释放，提高肠道内胰蛋白酶活性，增进氨基酸的代谢和蛋白质的合成[62]。在奥尼罗非鱼蛋氨酸缺乏的

实用饲料中补充晶体或包膜蛋氨酸,可显著提高罗非鱼生长性能和营养物质消化率,包膜蛋氨酸较晶体蛋氨酸具有更高的利用效率和更好的促生长效果[63]。黄鳝低蛋氨酸组在摄食后 6 h 出现蛋氨酸吸收峰值,晶体蛋氨酸与蛋氨酸羟基类似物钙盐组均在摄食后 9 h 出现蛋氨酸吸收峰值,包膜蛋氨酸组在摄食后 12 h 出现蛋氨酸吸收峰值,蛋氨酸羟基类似物组分别在摄食后 3 h 和 9 h 出现蛋氨酸峰值[64]。因此,晶体蛋氨酸包膜是提高蛋氨酸利用率的有效途径。

表 4-2 不同包被蛋氨酸饲料蛋氨酸溶失率（g/kg）

饲料蛋氨酸包被类型	10 min	30 min
羧甲基纤维素	23.4±4.7	44.8±3.9
玉米蛋白	29.1±1.6	41.5±0.5
κ-卡拉胶	21.6±2.6	43.0±11.2
琼脂	24.3±1.2	33.1±6.1
羧甲基纤维素-玉米蛋白	19.6±1.6	38.8±1.1
酪蛋白-明胶	21.4±3.0	30.0±1.8
玉米蛋白-κ-卡拉胶	20.8±5.4	46.6±0.4
酪蛋白-明胶-κ-卡拉胶	18.8±2.9	31.1±3.0
未包被	40.7±8.1	61.2±1.5

二、蛋氨酸羟基类似物及其钙盐的生物效价

黄鳝低鱼粉饲料中分别添加蛋氨酸有效含量为 0.2% 的晶体蛋氨酸、包膜蛋氨酸、蛋氨酸羟基类似物钙盐、蛋氨酸羟基类似物,蛋氨酸羟基类似物钙盐和蛋氨酸羟基类似物组显著提高增重率、蛋白质效率比、饲料转化率,提高肠道淀粉酶活力、血清葡萄糖和高密度脂蛋白胆固醇含量、肝脏谷草转氨酶活力,表明蛋氨酸羟基类似物钙盐和蛋氨酸羟基类似物优于其他形式蛋氨酸[64]。对建鲤的研究结果表明,在低蛋氨酸饲料中分别添加晶体蛋氨酸、微囊蛋氨酸、蛋氨酸羟基类似物及蛋氨酸羟基类似物钙盐,与未添加组相比,微囊蛋氨酸、蛋氨酸羟基类似物及蛋氨酸羟基类似物钙盐显著提高增重率和饲料转化率[65]。对斑点叉尾鮰的研究表明,在蛋氨酸缺乏饲料中分别添加晶体氨基酸、微囊蛋氨酸、蛋氨酸羟基类似物及蛋氨酸羟基类似物钙盐,微囊蛋氨酸、蛋氨酸羟基类似物钙盐对生长性能作用效果优于晶体氨基酸、蛋氨酸羟基类似物[66]。草鱼研究结果表明,蛋氨酸缺乏显著降低血清溶菌酶、酸性磷酸酶活性及补体 C3、补体 C4、免疫球蛋白 M 的含量;降低肝脏抗菌肽 2、头肾铁调素、β-防御素 1 基因的表达量,而添加 0.59%～0.62% 蛋氨酸羟基类似物则会显著提高上述指标,表明添加蛋氨酸羟基类似物具有提高鱼类抗菌能力的功能,且作用效果优于晶体蛋氨酸;依据生长性能,蛋氨酸羟基类似物的效价相当于蛋氨酸的 97%[67,68]。凡纳滨对虾在低鱼粉饲料中补充晶体蛋氨酸或羟基类似物钙盐,研究结果表明,与补充晶体蛋氨酸相比,低鱼粉饲料中补充羟基蛋氨酸钙能更有效地改善凡纳滨对虾的生长性能和饲料利用效率[69,70]。但对虹鳟的研究表明,在低鱼粉饲料中分别添加 DL-蛋氨酸、L-蛋氨酸、蛋氨酸羟基类似物钙盐对增重率、饲料转换效率、氮沉积效率等均没有显著影响[71]。因此,蛋氨酸羟基类似物钙盐、蛋氨酸羟基类似物在水产动

物中的应用效果优于蛋氨酸。

三、其他蛋氨酸的生物效价

还有其他形式蛋氨酸被应用到在水产动物饲料中。南美白对虾以增重率、特定生长率、饲料效率为指标,利用非线性指数回归模型分析显示,DL-甲硫氨酰基-DL-蛋氨酸的生物效价分别是 DL-蛋氨酸的 2.86、2.76 和 3.00 倍[72]。在对草鱼的研究中发现,DL-甲硫胺酰基-DL-蛋氨酸在生长性能和肠道健康方面作用效果优于蛋氨酸[73]。低鱼粉饲料中添加等量蛋氨酸寡肽比晶体蛋氨酸更能促进大黄鱼幼鱼、南美白对虾的生长及其对饲料的利用[74,75]。2-羟基-4-甲硫基丁酸作为一种有机酸在机体内酶系统的作用下可以转化为蛋氨酸,进而替代蛋氨酸为机体提供甲基,在大菱鲆的研究结果表明,蛋氨酸缺乏饲料中添加 2-羟基-4-甲硫基丁酸增重率、蛋白质沉积率、血清中抗坏血酸浓度均高于添加晶体蛋氨酸组[29]。

第六节 蛋氨酸对水产动物蛋白质代谢及健康的影响

一、蛋氨酸对机体蛋白质代谢的影响

蛋氨酸能够促进机体氨基酸平衡、转运、吸收及沉积,饲料中添加蛋氨酸显著提高肠道 γ-谷氨酰转肽酶(γ-GT)活性,进而促进氨基酸的吸收[10]。尖吻鲈摄食蛋氨酸缺乏和正常饲料后 2 h 血清中蛋氨酸含量达到峰值,蛋氨酸过量组摄食后 4 h 达到峰值,且摄食后 2~4 h 过量组苏氨酸、甘氨酸含量显著提高[76]。对大菱鲆的研究表明,蛋氨酸缺乏组降低血清中蛋氨酸、半胱氨酸、苏氨酸、精氨酸和组氨酸的含量,但甘氨酸、赖氨酸和丙氨酸含量显著升高[77]。蛋氨酸缺乏不仅影响血清氨基酸的含量,也会影响氨基酸在肌肉中的沉积率。对吉富罗非鱼的研究结果表明,蛋氨酸缺乏显著降低肌肉中蛋氨酸、缬氨酸、苏氨酸、精氨酸等,降低必需氨基酸总量、鲜味氨基酸总量及氨基酸总量[78]。蛋氨酸缺乏显著提高军曹鱼肌肉中苏氨酸、缬氨酸、异亮氨酸、精氨酸、组氨酸的含量,但对蛋氨酸、总氨基酸的含量影响没有达到显著水平[79]。虹鳟亲鱼饲料蛋氨酸缺乏降低鱼卵颗粒大小及鱼卵蛋氨酸、半胱氨酸、S-腺苷-L-甲硫氨酸、S-腺苷同型半胱氨酸的含量,进而降低孵化后仔鱼成活率[80]。异育银鲫研究表明,蛋氨酸缺乏或过量,血氨含量显著提高,表明氨基酸不平衡影响鱼类对氨基酸的利用[20]。上述研究表明,蛋氨酸缺乏或过量会导致机体氨基酸不平衡,影响其他氨基酸的吸收、转运、沉积,进而影响机体蛋白质的合成。

蛋氨酸影响机体蛋白质合成和肌肉品质,虹鳟蛋氨酸缺乏、正常及过量组白色肌肉总横截面积分别为 39.8、55.5、73.5 mm^2,白色肌肉的纤维总数分别为 $28.0×10^3$、$29.3×10^3$、$36.8×10^3$ 条/mm^2,主要原因是由于蛋氨酸含量的变化影响了肌肉生长及代谢相关基因的变化,蛋氨酸缺乏显著提高生肌因子 5(myogenic factor 5,$Myf5$)、成肌分化蛋白 1(Myogenic differentiation protein 1,$MyoD1$)、生肌调节因子 4(myogenic regulatory factors 4,$Mrf4$)、结合肌细胞增强因子(myocyte enhancer factor 2,$Mef2$)等基因表达,降低胶原蛋白 1α1(collagen 1α1,Col1α1)及胰岛素样生长因子-Ⅰ(IGF-Ⅰ)的表

达量[4]。蛋氨酸缺乏降低生长性能与降低机体蛋白质合成有关，主要是通过影响机体蛋白质合成相关信号通路基因的表达，进而降低机体蛋白质的合成。IGF-Ⅰ调控肌肉及其他组织的生长激素（growth hormone，GH）的分泌，GH进而调控机体合成代谢、生长及细胞分化。IGF-Ⅰ通过激活磷脂酰肌醇3-激酶/蛋白激酶B/雷帕霉素靶蛋白（PI3K/AKT/mTOR）信号通路调控机体蛋白质的合成，同时IGF-Ⅰ可以抑制肌肉蛋白质降解。蛋氨酸缺乏或过量降低大菱鲆脑垂体 GH 的基因表达量，降低肝脏生长激素受体（growth hormone receptor，GHR）、IGF-Ⅰ、雷帕霉素靶蛋白（TOR）、核糖体蛋白S6 激酶 1（Ribosomal protein S6 kinase 1，S6K1）的基因表达量[29]。蛋氨酸缺乏显著降低松浦镜鲤粗蛋白质含量、肌肉 IGF-Ⅰ、TOR、丝氨酸/苏氨酸蛋白激酶 11（serine-threonine kinase 11，STK11）、肌球蛋白重链（myosin heavy chain，MHC）的基因表达量[14]。对太平洋鲑的研究表明，蛋氨酸缺乏降低肌肉 IGF-Ⅰ、肌球蛋白轻重链（MLC）的基因表达量[81]。

二、蛋氨酸对肠道健康的影响

蛋氨酸缺乏或过量影响肠道形态结构、消化酶活性及吸收功能。蛋氨酸缺乏或过量均显著降低杂交石斑鱼前肠、中肠、后肠褶皱高度、褶皱宽度、肠道细胞高度及微绒毛高度[12]。草鱼中研究表明，蛋氨酸缺乏显著降低肠道褶皱高度，降低脂肪酶、蛋白酶、淀粉酶、γ-谷氨酰转肽酶（γ-glutamyl transpeptidase，γ-GT）、碱性磷酸酶、肌酸激酶的活性，同时降低脂肪酶、蛋白酶、淀粉酶的基因表达量[50]。大菱鲆研究发现，蛋氨酸缺乏会降低肠绒毛和微绒毛的高度，杯状细胞的数量及谷胱甘肽的含量，降低增殖细胞核抗原（proliferating cell nuclear antigen，PCNA）、黏蛋白2（mucoprotein-2，MUC2）基因的表达量[76]。

三、蛋氨酸对机体免疫、抗氧化功能的影响

蛋氨酸代谢生成的牛磺酸和谷胱甘肽，可清除体内氧自由基，参与机体免疫、抗氧化功能。欧洲鲈蛋氨酸缺乏显著降低肝脏 CAT 活性，提高谷胱甘肽还原酶和谷胱甘肽过氧化物酶活性[48]。蛋氨酸缺乏降低吉富罗非鱼补体 C3、补体 C4 的含量及溶菌酶、抗氧化相关酶的活性，提高丙二醛的含量[19,77]。蛋氨酸缺乏降低大黄鱼嗜水气单胞菌攻毒后成活率，显著降低血清溶菌酶活性、吞噬活性及总免疫球蛋白的含量，提高丙二醛的含量[18]。草鱼蛋氨酸缺乏组嗜水气单胞菌攻毒后出现皮肤出血和损伤，其主要可能机制是蛋氨酸缺乏激活 p38 丝裂原活化蛋白激酶/IκB 激酶 β/抑制性卡巴蛋白 α（p38MAPK/IKKβ/IκBα）信号通路，提高促炎因子的基因表达量；降低抗炎因子的基因表达量，进而促进炎症的发生；并通过 Kelch 样环氧氯丙烷相关蛋白 α/核因子 E2 相关因子 2（Keap1α/Nrf2）信号通路降低抗氧化基因表达量及活性，进而降低机体抗氧化能力和细胞完整性，导致肠炎发病率增加[67,73]。在嗜水气单胞菌攻毒条件下，露斯塔野鲮蛋氨酸含量为需要量 1.5 倍时，成活率显著高于蛋氨酸需要量组，但与需要量 2.0 倍组差异不显著[57]。上述研究表明，蛋氨酸缺乏或过量均影响水产动物的免疫和抗氧化功能，且在疾病或应激状态下蛋氨酸需要量更高，蛋氨酸过量对鱼类的毒性作用可能是由于肝脏 SAM 的积累造成的。

第七节 小结与展望

目前，关于蛋氨酸需要量在鱼类、虾蟹类的研究已经取得较大进展，鱼类对蛋氨酸需要量为0.73%~1.71%，虾蟹类对蛋氨酸的需要量为0.70%~1.28%。同时，关于蛋氨酸与牛磺酸、蛋氨酸与胆碱等其他营养素的相互作用也进行了大量研究，评价了不同形式蛋氨酸在水产动物中的应用效果及作用机制，开展了大量关于蛋氨酸缺乏或过量对机体蛋白质、脂肪、糖代谢、肠道健康及免疫功能的研究。

但是，关于蛋氨酸需要量的研究主要集中在幼鱼、幼虾阶段，关于生长中后期鱼虾的蛋氨酸需要量研究相对较少，有必要开展不同生长阶段的蛋氨酸需要量研究；蛋氨酸具有调控机体糖、脂代谢的功能，但需进一步探明其作用机制；蛋氨酸影响机体蛋白质代谢与合成，但关于蛋氨酸调控肌肉品质的研究相对较少，有必要开展蛋氨酸对肌肉品质影响的研究。

参考文献

[1] Obeid R. The metabolic burden of methyl donor deficiency with focus on the betaine homocysteine methyltransferase pathway [J]. Nutrients, 2013, 5: 3481-3495.

[2] Zhan X A, Li J X, Xu Z R, et al. Effects of methionine and betaine supplementation on growth performance, carcass composition and metabolism of lipids in male broilers [J]. British Poultry Science, 2006, 47: 576-580.

[3] Casero R A, Pegg A E. Polyamine catabolism and disease [J]. Biochemical Journal, 2009, 421: 323-338.

[4] Alami-durante H, Bazin D, Cluzeaud M, et al. Effect of dietary methionine level on muscle growth mechanisms in juvenile rainbow trout (*Oncorhynchus mykiss*) [J]. Aquaculture, 2018, 483: 273-285.

[5] Facanha F N, Oliveira-neto A R, Figueiredo-silva C, et al. Effect of shrimp stocking density and graded levels of dietary methionine over the growth performance of *Litopenaeus vannamei* reared in a green-water system [J]. Aquaculture, 2016, 463: 16-21.

[6] Facanha F N, Sabry-neto H, Figueiredo-silva C, et al. Minimum water exchange spares the requirement for dietary methionine for juvenile *Litopenaeus vannamei* reared under intensive outdoor conditions [J]. Aquaculture Research, 2018, 49: 1682-1689.

[7] 迟淑艳, 谭北平, 杨奇慧, 等. 投饲频率对摄食含有晶体蛋氨酸饲料的凡纳滨对虾生长性能的影响 [J]. 水产学报, 2013, 37 (5): 761-767.

[8] Nguyen M V, Espe M, Conceicao L E C, et al. The role of dietary methionine concentrations on growth, metabolism and N-retention in cobia (*Rachycentron canadum*) at elevated water temperatures [J]. Aquaculture Nutrition, 2019, 25: 495-507.

[9] He J Y, Tian L X, Lemme C, et al. The effect of dietary methionine concentrations on growth performance of juvenile Nile tilapia (*Oreochromis niloticus*) fed diets with two different digestible energy levels [J]. Aquaculture Nutrition, 2007, 23: 76-89.

[10] Wang X, Xue M, Figueiredo-silvac, et al. Dietary methionine requirement of the pre-adult gibel carp (*Carassius auratus gibeilo*) at a constant dietary cystine level [J]. Aquaculture Nutrition,

2016, 22 (3): 509-516.

[11] Chi S, He Y, Zhu Y, et al. Dietary methionine affects growth and the expression of key genes involved in hepatic lipogenesis and glucose metabolism in cobia (*Rachycentron canadum*) [J]. Aquaculture Nutrition, 2020, 26: 123-133.

[12] Li X J, Mu W, Wu X Y, et al. The optimum methionine requirement in diets of juvenile hybrid grouper (*Epinephelus fuscoguttatus* ♀ × *Epinephelus lanceolatus* ♂): Effects on survival, growth performance, gut micromorphology and immunity [J]. Aquaculture, 2020, 520: 735014.

[13] 杨炬懿, 王学武, 迟淑艳, 等. 蛋氨酸对2个规格斜带石斑鱼生长性能血清生化指标及肝脏酶活性的影响 [J]. 动物营养学报, 2020, 32 (3): 1305-1314.

[14] 程龙, 王连生, 徐奇友. 饲料蛋氨酸水平对松浦镜鲤生长性能、肌肉品质及肌肉合成通路相关基因表达的影响 [J]. 动物营养学报, 2020, 32 (3): 1293-1304.

[15] Yun Y H, Song D Y, He Z J, et al. Effects of methionine supplementation in plant protein based diet on growth performance and fillet quality of juveniles Yellow River carp (*Cyprinus carpio haematopterus*) [J]. Aquaculture, 2022, 549: 737810.

[16] 张庆功, 梁萌青, 徐后国, 等. 红鳍东方鲀幼鱼对饲料中蛋氨酸需求的研究 [J]. 渔业科学进展, 2019, 40 (4): 1-10.

[17] 段晶, 吴莉芳, 王婧瑶, 等. 蛋氨酸水平对洛氏鱥生长及消化酶和蛋白质代谢酶活力的影响 [J]. 西北农林科技大学学报（自然科学版), 2019, 47 (7): 23-31.

[18] Elmada C Z, Huang W, Jin M, et al. The effect of dietary methionine on growth, antioxidant capacity, innate immune response and disease resistance of juvenile yellow catfish (*Pelteobagrus fulvidraco*) [J]. Aquaculture Nutrition, 2016, 22: 1163-1173.

[19] He J Y, Long W Q, Han B, et al. Effect of dietary L-methionine concentrations on growth performance, serum immune and antioxidative responses of juvenile Nile tilapia, *Oreochromis niloticus* [J]. Aquaculture Research, 2017, 48: 665-674.

[20] Ren M, Liang H, He J, et al. Effects of DL-methionine supplementation on the success of fish meal replacement by plant proteins in practical diets for juvenile gibel carp (*Carassius auratus gibelio*) [J]. Aquaculture Nutrition, 2017, 23: 934-941.

[21] Wang Z, Mai K, Xu W, et al. Dietary methionine level influences growth and lipid metabolism via GCN2 pathway in cobia (*Rachycentron canadum*) [J]. Aquaculture, 2016, 454: 148-156.

[22] Wang Y Y, Che J F, Tang B B, et al. Dietary methionine requirement of juvenile *Pseudobagrus ussuriensis* [J]. Aquaculture Nutrition, 2016, 22: 1293-1300.

[23] Chu Z J, Gong Y, Lin Y C, et al. Optimal dietary methionine requirement of juvenile chinese sucker, *Myxocyprinus asiaticus* [J]. Aquaculture Nutrition, 2014, 20 (3): 253-264.

[24] Liao Y J, Ren M C, Liu B, et al. Dietary methionine requirement of juvenile blunt snout bream (*Megalobrama amblycephala*) at a constant dietary cystine level [J]. Aquaculture Nutrition, 2014, 20 (6): 741-752.

[25] Ahmed I. Dietary amino acid L-methionine requirement of fingerling Indian catfish, *Heteropneustes fossilis* (Bloch, 1974) estimated by growth and haemato-biochemical parameters [J]. Aquaculture Research, 2014, 45: 243-258.

[26] Elesho F E, Sutter D A H, Swinkels M A C, et al. Quantifying methionine requirement of juvenile African catfish (*Clarias gariepinus*) [J]. Aquaculture, 2021, 532: 736020.

[27] 向枭, 周兴华, 罗莉, 等. 饲料蛋氨酸水平对吉富罗非鱼生长饲料利用率和体成分的影响 [J]. 水

产学报, 2014, 38 (4): 537-548.

[28] Niu J, Du Q, Liu H Z, et al. Quantitative dietary methionine requirement of juvenile golden pompano, *Trachinotus ovatusr*, at a constant dietary cystine level [J]. Aquaculture Nutrition, 2013, 19 (5): 677-686.

[29] Ma R, Hou H, Mai K, et al. Comparative study on the effects of L-methionine or 2-hydroxy-4-(methylthio) butanoic acid as dietary methionine source on growth performance and anti-oxidative responses of turbot (*Psetta maxima*) [J]. Aquaculture, 2013, 412-413: 136-143.

[30] 贾鹏, 薛敏, 朱选, 等. 饲料蛋氨酸水平对异育银鲫幼鱼生长性能影响的研究 [J]. 水生生物学报, 2013, 37 (2): 217-226.

[31] 唐炳荣, 冯琳, 刘扬, 等. 生长中期草鱼蛋氨酸需要量的研究 [J]. 动物营养学报, 2012, 24 (11): 2263-2271.

[32] Ji K, Liang H L, Ge X P, et al. Optimal methionine supplementation improved the growth, hepatic protein synthesis and lipolysis of grass carp fry (*Ctenopharyngodon idella*) [J]. Aquaculture, 2022, 554: 738125.

[33] Zhou F, Xiao J X, HUA Y, et al. Dietary L-methionine requirement of juvenile black sea bream (*Sparus macrocephalus*) at a constant dietary cystine level [J]. Aquaculture Nutrition, 2011, 17: 469-481.

[34] Tulli F, Messina M, Calligaris M, et al. Response of European sea bass (*Dicentrarchus labrax*) to graded levels of methionine (total sulfur amino acids) in soya protein-based semi-purified diets [J]. British Journal o Nutrition, 2010, 104: 664-673.

[35] Zhou Q C, Wu Z H, Tan B P, et al, Optimal dietary methionine requirement for Juvenile Cobia (*Rachycentron canadum*) [J]. Aquaculture, 2016, 258: 551-557.

[36] Luo Z, Liu Y, Mai K, et al, Dietary L-methionine requirement of juvenile grouper *Epinephelus coioides* at a constant dietary cystine level [J]. Aquaculture, 2015, 249: 409-418.

[37] Wang L, Ye L, Hua Y, et al. Effects of dietary DL-methionyl-DL-methionine (Met-Met) on growth performance, body composition and haematological parameters of white shrimp (*Litopenaeus vannamei*) fed with plant protein-based diets [J]. Aquaculture Research, 2019, 50: 1718-1730.

[38] 朱杰, 蒋广震, 徐维娜, 等. 日本沼虾的蛋氨酸需要量 [J]. 经济动物学报, 2014, 18 (3): 151-158.

[39] 朱杰, 徐维娜, 张微微, 等. 克氏原螯虾的适宜蛋氨酸需要量 [J]. 中国水产科学, 2014, 21 (2): 300-309.

[40] 周小秋, 杨凤, 周安国, 等. 稚鳖蛋氨酸的营养需要量 [J]. 水生生物学报, 2003, 27 (1): 69-73.

[41] 叶金云, 王友慧, 郭建林, 等. 中华绒螯蟹对赖氨酸、蛋氨酸、精氨酸的需要量 [J]. 水产学报, 2010, 34 (10): 1541-1548.

[42] 金敏, 王猛强, 霍雅文, 等. 三疣梭子蟹幼蟹的蛋氨酸需要量 [J]. 动物营养学报, 2015, 27 (11): 3457-3467.

[43] Candebat C L, Both M, Codabaccus M B, et al. Dietary methionine spares the requirement for taurine in juvenile Yellowtail Kingfish (*Seriola lalandi*) [J]. Aquaculture, 2020, 522: 735090.

[44] Gayldrd T G, Barrows F T, Teague A M, et al. Supplementation of taurine and methionine to all-plant protein diets for rainbow trout (*Oncorhynchus mykiss*) [J]. Aquaculture, 2007, 269: 514-524.

[45] Michelato M, Furuy W M, Gatlin Ⅲ D M, et al. Metabolic responses of Nile tilapia *Oreochromis niloticus* to methionine and taurine supplementation [J]. Aquaculture, 2018, 485: 66-72.

[46] De Moura L B, Diogenes A F, Campelo D A. V, et al. Taurine and methionine supplementation as a nutritional strategy for growth promotion of meagre (*Argyrosomus regius*) fed high plant protein diets [J]. Aquaculture, 2018, 497: 389-395.

[47] De Moura L B, Diogenes A F, Campelo D A V, et al. Nutrient digestibility, digestive enzymes activity, bile drainage alterations and plasma metabolites of meagre (*Argyrosomus regius*) feed high plant protein diets supplemented with taurine and methionine [J]. Aquaculture, 2019, 511: 1-7.

[48] Coutinho F, Simoes R, Mongeortiz R, et al. Effects of dietary methionine and taurine supplementation to low-fish meal diets on growth performance and oxidative status of European sea bass (*Dicentrarchus labrax*) juveniles [J]. Aquaculture, 2017, 479: 447-454.

[49] Poppi D A, Moors S S, Glencross B D. The effect of taurine supplementation to a plant-based diet for barramundi (*Lates calcarifer*) with varying methionine content [J]. Aquaculture Nutrition, 2018, 24: 1340-1350.

[50] Jiang J, Shi D, Zhou X Q, et al. Effects of lysine and methionine supplementation on growth, body composition and digestive function of grass carp (*Ctenopharyngodon idella*) fed plant protein diets using high-level canola meal. Aquaculture Nutrition, 2016, 22: 1126-1133.

[51] 陆静, 付文忠, 邵庆均. 鱼粉部分替代后添加赖氨酸和蛋氨酸对黑鲷消化作用的影响 [J]. 扬州大学学报 (农业与生命科学版), 2014, 35 (2): 34-37.

[52] Niu J, Figueiredo-Silvac, Dong Y, et al. Effect of replacing fish meal with soybean meal and of DL-methionine or lysine supplementation in pelleted diets on growth and nutrient utilization of juvenile golden pompano (*Trachinotus ovatus*) [J]. Aquaculture Nutrition, 2016, 22 (3): 606-614.

[53] Zhang Y Q, Ji W X, Wu Y B, et al. Replacement of dietary fish meal by soybean meal supplemented with crystalline methionine for Japanese seabass (*Lateolabrax japonicus*) [J]. Aquaculture Research, 2016, 47: 243-252.

[54] Espe M, Andersen S M, Holen E, et al. Methionine deficiency does not increase polyamine turnover through depletion of liver S-adenosylmethionine (SAM) in juvenile Atlantic salmon [J]. British Journal of Nutrition, 2014, 112: 1274-1283.

[55] Espe M, Zerrahn J E, Holen E, et al. Choline supplementation to low methionine diets increase phospholipids in Atlantic salmon, while taurine supplementation had no effects on phospholipid status, but improved taurine status [J]. Aquaculture Nutrition, 2016, 22: 776-785.

[56] Michael F R, Koshio S, Techima S, et al. Effect of choline and methionine as methyl group donors on juvenile kuruma shrimp, *Marsupenaeus japonicus* Bate [J]. Aquaculture, 2006, 258: 521-528.

[57] Mira I N, Sahu N P, Pal A K, et al. Synergistic effect of L-methionine and fucoidan rich extract in eliciting growth and non-specific immune response of *Labeo rohita* fingerlings against *Aeromonas hydrophila* [J]. Aquaculture, 2017, 479: 396-403.

[58] Chu Z J, Yu D H, Dong G F, et al. Partial replacement of fish meal by soybean meal with or without methionine and phytase supplement in diets for juvenile Chinese sucker, *Myxocyprinus asiaticus* [J]. Aquaculture Nutrition, 2016, 22: 989-996.

[59] Espe M, Vikesav helgoy T T, Adam A C, et al. Atlantic salmon fed a nutrient package of surplus methionine, vitamin B_{12}, folic acid and vitamin B_6 improved growth and reduced the relative liver size, but when in excess growth reduced [J]. Aquaculture Nutrition, 2020, 26: 477-489.

[60] Alam M S, Teshima S, Koshio S, et al. Effects of supplementation of coated crystalline amino acids on growth performance and body composition of juvenile kuruma shrimp *Marsupenaeus japonicus* [J].

Aquaculture Nutrition, 2004, 10: 309-316.

[61] Alam M S, Teshima S, Koshio S, et al. Supplemental effects of coated methionine and/or lysine to soy protein isolate diet for juvenile kuruma shrimp, *Marsupenaeus japonicas* [J]. Aquaculture, 2005, 248: 13-19.

[62] 迟淑艳, 谭北平, 董晓慧, 等. 微胶囊蛋氨酸或晶体蛋氨酸对军曹鱼幼鱼相关酶活性的影响 [J]. 中国水产科学, 2011, 18 (1): 110-118.

[63] 冷向军, 田娟, 陈丙爱, 等. 罗非鱼对晶体蛋氨酸、包膜蛋氨酸利用的比较研究 [J]. 水生生物学报, 2013, 37 (2): 235-242.

[64] 胡亚军, 胡毅, 石勇, 等. 不同形式蛋氨酸对黄鳝生长、血清生化、血清游离氨基酸含量及肌肉品质的影响 [J]. 水生生物学报, 2019, 43 (6): 1155-1163.

[65] 单玲玲, 李小勤, 郑小森, 等. 不同形式蛋氨酸对建鲤生长性能及血清游离氨基酸含量的影响 [J]. 水生生物学报, 2015, 39 (2): 259-266.

[66] Zhao J X, Li X Q, Leng X J, et al. Comparative study on the utilization of different methionine sources by channel catfish, *Ictalurus punctatus* (Rafinesque, 1818) [J]. Aquaculture Research, 2017, 48: 3618-3630.

[67] Pan F Y, Feng L, Ling W D, et al. Methionine hydroxy analogue enhanced fish immunity via modulation of NF-kB, TOR, MLCK, MAPKs and Nrf2 signaling in young grass carp (*Ctenopharyngodon idella*) [J]. Fish & Shellfish Immunology, 2016, 56: 208-228.

[68] Wu P, Pan F Y, Feng L, et al. Methionine hydroxy analogue supplementation modulates gill immunological and barrier health status of grass carp (*Ctenopharyngodon idella*) [J]. Fish and Shellfish Immunology, 2018, 74: 637-648.

[69] 黄文文, 霍雅文, 王猛强, 等. 低鱼粉饲料中补充晶体蛋氨酸和羟基蛋氨酸钙在凡纳滨对虾上饲喂效果的比较研究 [J]. 动物营养学报, 2015, 27 (6): 1722-1732.

[70] Chen J, Li X, Huan D, et al. Comparative study on the utilization of crystalline methionine and methionine hydroxy analogue calcium by Pacific white shrimp (*Litopenaeus vannamei* Boone) [J]. Aquaculture Research, 2018, 49: 3088-3096.

[71] Powell C D, Chowdhury M A K, Bureau D P. Assessing the bioavailability of L-methionine and a methionine hydroxy analogue (MHA-Ca) compared to DL-methionine in rainbow trout (*Oncorhynchus mykiss*) [J]. Aquaculture Research, 2017, 48: 332-346.

[72] Niu J, Lemme A, He J Y, et al. Assessing the bioavailability of the Novel Met-Met product (AQUAVI® Met-Met) compared to DL-methionine (DL-Met) in white shrimp (*Litopenaeus vannamei*) [j]. Aquaculture, 2018, 484: 322-332.

[73] Su Y N, Wu P, Feng L, et al. The improved growth performance and enhanced immune function by DL methionyl-DL-methionine are associated with NF-κB and TOR signalling in intestine of juvenile grass carp (*Ctenopharyngodon idella*) [J]. Fish and Shellfish Immunology, 2018, 74: 101-118.

[74] 马俊, 魏泽宏, 邢淑娟, 等. 饲料中添加蛋氨酸寡肽对大黄鱼 (*Larimichthys crocea*) 生长、饲料利用和蛋白质代谢反应的影响 [J]. 渔业科学进展, 2016, 37 (3): 126-133.

[75] Gu M, Zhang W B, Bai N, et al. Effects of dietary crystalline methionine (CMet) or oligo-methionine (OMet) on growth performance and feed utilization of white shrimp, *Litopenaeus vannamei* fed plant protein-enriched diets [J]. Aquaculture Nutrition, 2013, 19: 39-46.

[76] Poppi D A, Moore S S, Wade N M, et al. Postprandial plasma free amino acid profile and hepatic

gene expression in juvenile barramundi (*Lates calcarifer*) is more responsive to feed consumption than to dietary methionine inclusion [J]. Aquaculture, 2019, 501: 345-358.

[77] Gao Z Y, Wang X, Tan C, et al. Effect of dietary methionine levels on growth performance, amino acid metabolism and intestinal homeostasis in turbot (*Scophthalmus maximus* L.) [J]. Aquaculture, 2018, 498: 335-342.

[78] 向枭, 周兴华, 曾本和, 等. 蛋氨酸水平对吉富罗非鱼肌肉氨基酸组成及血清抗氧化能力的影响 [J]. 水产学报, 2016, 40 (6): 1359-1367.

[79] 何远法, 郭勇, 迟淑艳, 等. 低鱼粉饲料中补充蛋氨酸对军曹鱼生长性能体成分及肌肉氨基酸组成的影响 [J]. 动物营养学报, 2018, 30 (2): 624-634.

[80] Fontagne-dicharry S, Alami-durante H, Aragao C, et al. Parental and early-feeding effects of dietary methionine in rainbow trout (*Oncorhynchus mykiss*) [J]. Aquaculture, 2017, 469: 16-27.

[81] Espe M, Veiseth-kent E, Zerrahn J E, et al. Juvenile atlantic salmon decrease white trunk muscle igf-1 expression and reduce muscle and plasma free sulphur amino acids when methionine availability is low while liver sulphur metabolites mostly is unaffected by treatment [J]. Aquaculture Nutrition, 2016, 22 (4): 801-812.

第五章 苏氨酸

苏氨酸是动物体内代谢途径中唯一不经过脱氨基作用和转氨基作用的氨基酸。苏氨酸作为水产动物生长所需的必需氨基酸，在促进机体生长、强化体蛋白质合成及增强机体免疫力等方面扮演着重要角色。本章主要对水产动物苏氨酸营养需要，苏氨酸对肉质、蛋白质代谢、消化吸收、抗氧化能力、免疫功能的影响进行了综述，以期为苏氨酸在水产动物营养免疫与健康养殖方面的进一步研究奠定理论基础。

第一节 苏氨酸的理化性质

苏氨酸（Threonine，Thr），分子式 $C_4H_9NO_3$，分子质量 119.12，熔点 253 ℃，分子结构如图 5-1 所示，为白色斜方晶系或结晶性粉末。无臭，味微甜。不溶于乙醇、乙醚和氯仿。

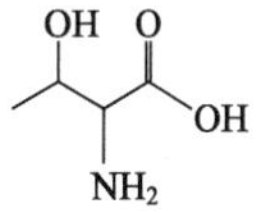

图 5-1 苏氨酸分子结构

第二节 水产动物苏氨酸需要量

苏氨酸的缺乏可以导致日本牙鲆[1]和印度鲮[2]生长减缓，露斯塔野鲮[3]和印度鲇[11]饲料利用率降低。近 20 年发表的水产动物苏氨酸的需要量研究见表 5-1。由表 5-1 可以看出，以占饲料的百分比计算，鱼类对苏氨酸的需要量为 1.04%～2.06%，虾蟹类对苏氨酸的需要量为 1.36%～1.59%。不同种类水产动物的苏氨酸需要量存在差异，虹鳟的苏氨酸需要量仅为 1.04%[5]，而大黄鱼的苏氨酸需要量高达 2.06%[6]。同一品种不同规格的水产动物对苏氨酸需要量的变化情况存在差异，体重为 8 g[6]、333.93 g[7]鲈的苏氨酸需要量分别为 1.78%、1.84%，二者相差不大，而体重 8.35 g[8]、250.00 g[9]草鱼的苏氨酸需要量分别为 1.72%、1.22%，二者相差较大。因此，规格对于水产动物苏氨酸需要量的影响规律尚未清晰。不同养殖环境对水产动物苏氨酸需要量潜在影响，Huai 等[15]在低盐度条件下得出的 0.48 g 凡纳滨对虾苏氨酸的需要量为 1.36%。而王用黎[16]发现在正常海水盐度下 0.50 g 凡纳滨对虾苏氨酸需要量为 1.51%。在目前的研究中，生长参数通常是评价水产动物苏氨酸需要量的常用指标，但需指出的是不同评价指标会对水产动物苏氨酸需要量产生一定影响，如大黄鱼分别以特定生长率、氮保持率作为评价指标，回归分析得到的苏氨酸需要量为 1.86%、2.06%。根据免疫指标确定的幼建鲤苏氨酸需要量为 1.56%～1.69%，与根据生长性能指标确定的 1.62%以及根据消化吸收能力性能指标确定的 1.62%~1.69%存在一定差异[10]。

表 5-1 水产动物苏氨酸需要量

	品种	体重（g）	模型	指标	需要量（占饲料干物质,%）	参考文献
鱼类	日本牙鲆	2.00	折线模型	增重率	1.61	Alam 等[1]
	露斯塔野鲮	0.58	二次多项式模型	增重率	1.55	Abidi 等[3]
				蛋白质效率	1.72	
	印度鲮	3.85	二次多项式模型	饲料系数	1.81	Ahmed 等[2]
				蛋白质效率	1.78	
	印度鲇	3.60	二次多项式模型	饲料效率	1.27	Ahmed 等[4]
	虹鳟	1.80	折线模型	最佳苏氨酸沉积	1.04	Bodin 等[5]
	鲈	8.00	二次多项式模型	特定生长率	1.78	何志刚[6]
				氮保持率	1.87	
		333.93	二次多项式模型	特定生长率	1.84	窦秀丽[7]
				饲料效率	1.87	
				蛋白质沉积	1.83	
	草鱼	8.35	折线模型	特定生长率	1.72	文华等[8]
				饲料系数	1.66	
				蛋白质沉积率	1.61	
		250.00	折线模型	增重率	1.22	胡晓霞[9]
	大黄鱼	6.00	二次多项式模型	特定生长率	1.86	何志刚[6]
				氮保持率	2.06	
	建鲤	18.00	二次多项式模型	增重率	1.62	冯琳[10]
				特定生长率	1.62	
				饲料效率	1.69	
				血清溶菌酶	1.57	
				白细胞数量	1.63	
				免疫球蛋白	1.69	
				肠道胰蛋白酶	1.65	
				肠道脂肪酶	1.62	
	尼罗罗非鱼	37.61	折线模型	增重率	1.35	Silva 等[11]
	吉富罗非鱼	67.08	二次多项式模型	增重率	1.72	周兴华等[12]
				饲料系数	1.62	
				蛋白质效率	1.62	
	大西洋鲑	79.00	指数模型	苏氨酸获得率	1.31	Helland 等[13]
甲壳类	斑节对虾	0.05	二次多项式模型	增重率	1.40	Millamena 等[14]
	凡纳滨对虾（低盐度条件）	0.48	折线模型	特定生长率	1.36	Huai 等[15]
	凡纳滨对虾（正常海水环境）	0.50	折线模型	特定生长率	1.51	王用黎[16]
	中华绒螯蟹	0.36	折线模型	增重率	1.59	王伟等[17]

需要指出的是由于在试验设计、鱼种及养殖条件等方面存在差异，表5-1中列出的研究结果在具有合理性的同时又存在差异性，造成分析过程具有主观性，分析结果多为相对模糊的定性结论。为了避免这一问题，采用Meta分析方法对水产动物苏氨酸需要研究中的原始结果进行了二次整合的定量分析[18]。初检出相关文献20篇，经筛选最终选择14篇，其中关于鱼类的相关研究11篇，甲壳动物3篇（对虾2篇、中华绒螯蟹1篇）。基于分析剂量效应，使用四参数营养动力学分析方法对经过标准化响应处理后的增重率（Weight gain rate，WGR）与苏氨酸需要量进行定量分析[19]。根据不同鱼类、甲壳类对苏氨酸需要量的研究结果，分别采用了WGR与苏氨酸占饲料百分比（图5-2A）和占饲料中粗蛋白百分比（图5-2B）绘制曲线，并尝试采用不同回归分析对曲线进行拟合，最终确定了三次多项式回归分析模型（拟合曲线公式分别为：$y=20.923x^3-105.41x^2+176.08x-0.0084$，$R^2=0.9957$；$y=1.5556x^3-18.84x^2+74.725x+0.0129$，$R^2=0.9798$），拟合得到鱼类和甲壳类水产动物苏氨酸需要量为1.572%（占饲料）和4.673%（占饲料蛋白）。

此外，饲料蛋白源的差异也会直接影响水产动物对苏氨酸的需要。在确定水产动物苏氨酸营养需要的试验中，通常会选取如豆粕、玉米蛋白、小麦面筋等苏氨酸含量较低的植物蛋白源，因此在饲料中以上述蛋白源为鱼体提供蛋白时应考虑饲料中苏氨酸含量是否能够满足水产动物的营养需[20]。王亚玲等[21]发现以特定生长率、蛋白质效率为评价指标，用低鱼粉（15%）饲料饲喂三倍体虹鳟时，其苏氨酸最佳需要量为1.2%～1.3%。

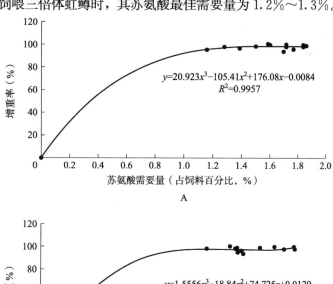

图5-2 不同鱼类苏氨酸需要量的文献分析

第三节　苏氨酸对水产动物肉质的影响

肉质评价的内容主要包括外观特征、肌肉营养生化组成、感官评价结合质构特性和肌纤维组织特征等方面。肌肉的常规营养成分含量是营养物质摄入后在体内沉积的一个具体表现，可以评估肌肉的食用营养价值。Alam 等[1]发现饲料中苏氨酸对日本牙鲆全鱼营养成分均有显著影响，当苏氨酸水平达到需要量 1.61%（占饲料），全鱼水分最低，蛋白含量最高，同时苏氨酸水平的提高也会显著提升全鱼粗脂肪含量。在对印度鲇[4]和幼龄草鱼[13]研究中也得到了类似结论。这表明适宜苏氨酸水平能够提高机体蛋白质和脂肪的沉积能力。在对生长中期草鱼的研究中同样发现，饲料中苏氨酸水平达到 1.51% 时，能够降低草鱼肌肉水分含量，增加肌肉粗蛋白质含量[9,22]。肌肉质构特性与肌肉微观组织结构密切关联，鱼肉质构特性的变化主要取决于鱼肉肌纤维数量和结构上的改变[23]。苏氨酸过量或缺乏均会扩大建鲤肌纤维平均直径，缩小肌纤维密度，进而降低肌肉肌纤维募集能力，减缓肌肉生长速度，同时反映出肌肉硬度小、剪切力差[24]。而对于凡纳滨对虾[16]、中华绒螯蟹[17]，肌肉营养成分并未受到饲料苏氨酸水平的显著影响。苏氨酸在调控鱼类和虾蟹类肉质方面的不同机制，可能是造成二者饲料苏氨酸需要量差异的重要原因之一。

第四节　苏氨酸对水产动物蛋白质代谢的影响

对于幼建鲤，当苏氨酸水平低于 1.57% 或高于 1.57% 时均会降低肝胰脏和肌肉谷丙转氨酶和谷草转氨酶的活力，提高二者在血清中的活力，并显著提升血清中血氨的含量[10]，表明苏氨酸水平过低或过高均会影响机体氨基酸平衡，使蛋白质分解代谢处于优势，抑制机体蛋白质合成代谢，暗示饲料中苏氨酸水平的变化会影响鱼体蛋白质的周转代谢[25]。RNA/DNA 比值能够精准地反映蛋白质合成情况[26]。苏氨酸缺乏会降低草鱼肌肉 RNA 含量以及 RNA/DNA 比值，表明苏氨酸缺乏会减缓肌肉蛋白质合成速度[8,27]。雷帕霉素蛋白是蛋白质合成的信号调控分子[28]。不同于亮氨酸[29]、异亮氨酸[30]及缬氨酸[31]，当苏氨酸缺乏或过量时幼建鲤的肌肉和肠道中 TOR 基因表达量会被显著上调[32]，这可能与前述三者均为支链氨基酸有关，支链氨基酸调控蛋白质最主要的潜在机制是通过激活 mTOR 信号通路来加速肌肉蛋白质合成[33]。因此，关于苏氨酸影响水产动物蛋白质代谢的具体机制需要进一步研究。

第五节　苏氨酸对水产动物消化吸收能力的影响

肝胰脏、肠道的生长发育程度影响着水产动物的消化吸收[34]。对于幼建鲤[10]及生长后期草鱼[35]，苏氨酸缺乏会显著降低肠体指数、肠长指数、肠道蛋白含量、肝体指数及肝胰脏蛋白含量。同时，对体外培养鲤肠上皮细胞的研究发现，适宜水平的苏氨酸能够促进鲤肠上皮细胞的增殖和分化[36]，表明适宜的苏氨酸水平能够促进鱼类肠道及肝胰脏的生长发育。肠道及肝胰脏中消化酶活力能够直接反映水产动物消化能力水平[37]。对于生长中期草鱼[9]、幼建鲤[10]、生长后期草鱼[34]，饲料苏氨酸缺乏将会抑制肠道和肝胰脏中蛋白酶、脂肪酶和淀粉酶活力，而添加苏氨酸后，三种酶的合成会显著增加，在王伟等[17]对中华绒螯蟹的研究中得到了一致的结果。同时，这在赖氨酸[38]、蛋氨酸[39]等对水产动物消化酶影响研究中也发现了类似的结果，说明当机体所需的氨基酸不足时，外源补充适量的氨基酸能够促进消化酶的合成和分泌。肠道的吸收能力主要依靠刷状缘酶的活力[40]。饲料苏氨酸缺乏后，生长中期草鱼肠道中碱性磷酸酶（alkaline phosphatase, ALP）、肌酸激酶（creatine kinase, CK）、γ-谷氨酰转肽酶和 Na^+/K^+ -ATP 酶活力显著降低，而当苏氨酸水平添加适宜时，肠道中四种酶的活力显著提升[9]。Habte - Tsion 等[41]发现苏氨酸除了提高团头鲂肠道各段 ALP、CK、γ-GT 和 Na^+/K^+ -ATP 酶活力外，还会显著上调四种刷状缘酶的基因表达水平，进一步从分子水平揭示了苏氨酸促进水产动物消化吸收能力的潜在机制。

第六节 苏氨酸对水产动物抗氧化能力的影响

当过量产生活性氧簇（Rreactive oxygen species，ROS）或体内酶性、非酶性抗氧化系统不足以清除时，则会导致氧化应激，丙二醛（MDA）含量能够反映机体和组织脂质过氧化程度[42]。苏氨酸缺乏会显著增加生长中期草鱼血清、肝胰脏和肌肉 MDA 含量，而添加适宜苏氨酸后其含量则会显著降低，血清和肝胰脏 MDA 含量在1.26%苏氨酸水平时达到最小值，肌肉 MDA 含量则在0.97%和1.26%苏氨酸水平时达到最小值[16]，表明适宜苏氨酸水平能够抑制水产动物机体和组织脂质过氧化的程度。鱼体应对氧化应激主要依赖于抗氧化系统，包括非酶抗氧化物质及抗氧化酶[42]。与生长中期草鱼组织中 MDA 含量变化相对应，苏氨酸缺乏会导致血清、肝胰脏和肌肉超氧化物歧化酶（SOD）、谷胱甘肽过氧化物酶（GPx）、过氧化氢酶（CAT）、谷胱甘肽转移酶（GST）、谷胱甘肽还原酶（GR）和谷胱甘肽（GSH）活力的降低[9]，而添加适宜苏氨酸后上述抗氧化酶活力则会显著提升。Habte - Tsion 等[43]发现苏氨酸缺乏会显著降低团头鲂肝胰脏 SOD、CAT 和 GPx 活力，而当苏氨酸水平在1.58%~2.08%时，肝胰脏中三种抗氧化酶的活力显著提升。鱼体抗氧化能力主要受到 Kelch 样环氧氯丙烷相关蛋白 α/核因子 E2 相关因子 2（Kelch - like epichlorohydrin associated protein α/nuclear factor E2 associated factor 2, Keaplα/Nrf2）信号通路调控，且近些年的研究已经表明饲料中亮氨酸[44]、蛋氨酸[45]等必需氨基酸的水平对此通路有着明显的调控作用。对于团头鲂，苏氨酸缺乏会显著下调肝胰脏中 $Nrf2$ 的基因表达量及下游相关抗氧化基因，如铜锌超氧化物歧化酶（Cu/Zn - SOD）、锰超氧化物歧化酶（Mn - SOD）、CAT、GPx、GST，当苏氨酸水平达到1.58%时，上述抗氧化相关基因表达量均达到峰值，揭示苏氨酸改善机体抗氧化能力的可能机制同样是调控 Nrf2 信号通路。目前，关于苏氨酸对虾蟹类抗氧化能力影响的文章还较少，仅见于苏氨酸对中华绒螯蟹肝胰腺 SOD 活力影响的研究，苏氨酸水平超过1.61%时肝胰腺 SOD 活力显著降低，进而可能诱发氧化损伤。因此，还需进行更加深入的研究探讨苏氨酸影响虾蟹类抗氧化能力的作用机制。

第七节　苏氨酸对水产动物免疫功能的影响

苏氨酸对水产动物免疫功能的影响可以从特异性免疫和非特异性免疫两个方面进行探讨。

苏氨酸缺乏或过量影响水产动物特异性免疫功能的可能原因有以下几个方面：第一，减缓免疫器官的发育。苏氨酸缺乏会显著降低幼建鲤头肾重量、脾脏重量及相应体指数[10]。第二，降低免疫球蛋白含量。免疫球蛋白M（immunoglobulin M，IgM）可能是鱼体内仅有的一类免疫球蛋白[46]。对于幼建鲤，苏氨酸水平由1.57%降至0.74%，血清中IgM含量会下降28.8%，进而削弱幼建鲤特异性免疫能力。而对于瓦氏黄颡鱼（Pelteobagrus vachelli），苏氨酸水平由1.40%降至0.28%，血清中IgM和白蛋白含量未出现显著变化，说明不同鱼类血清中IgM对苏氨酸的敏感水平并不相同[47]。第三，抑制补体系统的激活。补体3（C3）、补体4（C4）是增强鱼体特异性免疫应答的主要因子。苏氨酸缺乏或过量显著降低幼建鲤血清中补体C3、补体C4含量[10]，在Dong等[49]对苏氨酸影响草鱼鳃中C3、C4含量的研究中的得到了一致的结果。第四，减少血液中红细胞数量。红细胞可以诱导静态B细胞增殖分化为产生抗体（免疫球蛋白）的浆细胞[48]。当苏氨酸水平由1.57%降至0.74%时，幼建鲤血液中红细胞数量下降30.9%[3]；而当苏氨酸水平低于2.08%或高于2.08%时，团头鲂血液中红细胞数量则会显著下降[44]。

苏氨酸缺乏或过量主要通过降低非特异性免疫相关酶活性、提高促炎因子的基因表达量影响鱼体非特异性免疫功能。苏氨酸缺乏或过量显著降低幼建鲤血清中溶菌酶、酸性磷酸酶（acid phosphatase，ACP）活力，以血清溶菌酶和酸性磷酸酶为指标，进行二次多项式回归分析得到幼建鲤苏氨酸需要量分别为1.57%、1.60%[10]。苏氨酸过量会显著提升团头鲂肝胰脏及前肠促炎因子肿瘤坏死因子-α（tumor necrosis factor-α，TNF-α）基因表达量，苏氨酸缺乏或过量则会显著提高中肠TNF-α基因表达量，表明苏氨酸无论是缺乏还是过量都会引发机体炎症反应，影响机体健康[49]。作为鱼体呼吸和氨氮排泄的主要组织器官，鳃通过形成免疫屏障和物理屏障增强对病原体入侵的抵抗力[50]。苏氨酸缺乏会提升团头鲂柱状黄杆菌攻毒后烂鳃病的发病率，通过减少抗菌物质分泌、上调促炎因子基因表达及下调抗炎因子的基因表达等方式破坏鳃的免疫屏障功能，通过降低鳃抗氧化能力、促使细胞凋亡及破坏紧密连接等方式削弱鳃的物理屏障功能。苏氨酸缺乏会显著下调草鱼肠道中信号分子核因子κB（NF-κB）、TOR、Nrf2、c-Jun氨基末端激酶（c-Jun amino-terminal kinase，JNK）和肌球蛋白轻链激酶（myosin light chain kinase，MLCK）基因表达量，进而导致鱼类免疫和物理屏障功能恶化[51]。

目前，关于苏氨酸对虾蟹类免疫功能影响的研究还较少，仅见于苏氨酸对中华绒螯蟹肝胰腺ACP和ALP活力影响的研究，苏氨酸缺乏会显著降低肝胰腺中两种免疫相关酶的活力，降低机体非特性免疫能力，进而影响机体生长发育，但苏氨酸影响虾蟹类抗氧化能力的作用机制仍需进一步深入探讨。

第八节 小结与展望

　　近年来,关于水产动物苏氨酸营养的研究逐渐增多,发现了鱼类苏氨酸的需要量在 1.04%～2.06%,虾蟹类苏氨酸需要量在 1.36%～1.59%;初步明晰了苏氨酸对水产动物肉质、蛋白质代谢、消化吸收功能、抗氧化功能及免疫功能的影响。未来对于水产动物苏氨酸营养的研究应从以下四个方面深入拓展:目前苏氨酸营养需要研究主要集中于鱼类,关于虾蟹类的研究相对较少,且仅有的研究局限于幼体阶段,需开展至生长中后期;以苏氨酸调控肌肉品质的具体机制作为切入点,研究鱼类及虾蟹类对苏氨酸营养需要差异的潜在原因;作为蛋白质合成的重要氨基酸的最主要限制性氨基酸之一,苏氨酸调控水产动物体蛋白合成和鱼类免疫球蛋白生成的具体作用机制值得进行深入研究;新型蛋白源开发及高植物蛋白添加成为低鱼粉饲料研制的主要方式,但因水产动物对蛋白源的吸收利用存在差异,进而造成对饲料氨基酸需要量的不同,因此未来应加强低鱼粉饲料开发及氨基酸需要的同步研究。上述研究内容将为更深入全面地了解苏氨酸在水产动物上的营养免疫功能及其作用机制提供新思路新方向,为水产动物健康养殖及绿色低氮饲料开发奠定坚实理论基础。

参考文献

[1] Alam M, Teshima S, Koshio S, et al. Optimum dietary threonine level for juvenile Japanese flounder *Paralichthys olivaceus* [J]. Asian Fisheries Science, 2013, 16 (1/2): 175-184.

[2] Ahemd L, Khan M A, Jafri A K. Dietary threonine requirement of fingerling Indian major carp, *Cirrhinus mriigala* (Hamilton) [J]. Aquaculture Research, 2004, 35 (2): 162-170.

[3] Abidi S F, Khan M A. Dietary threonine requirement of fingerling Indian major carp, *Laheo rohita* (Hamilton) [J]. Aquaculture Research, 2008, 39 (14): 1498-1505.

[4] Ahemd I. Dietary amino acid I – threonine requirement of fingerling Indian catfish, *Heteropneuste fossilis* (Block) estimated by growth and biochemical parameters [J]. Aquaculture International, 2007, 15 (5): 337-350.

[5] Bodin N, Mambrini M, Wauters J B, et al. Threonine requirements for rainbow trout (*Oncohynchus mykiss*) and Atlantic salmon (*Salmo salar*) at the fry stage are similar [J]. Aquaculture, 2008, 274 (2-4): 353-355.

[6] 何志刚. 大黄鱼(*Pseudosciaena crocea* R.)和鲈鱼(*Lateolabrax japonicas*)苏氨酸和苯丙氨酸营养生理研究 [D]. 青岛:中国海洋大学, 2008.

[7] 窦秀丽,梁萌青,郑珂珂,等. 鲈鱼(*Lateolabrax japonicus*)生长后期对苏氨酸需要量 [J]. 渔业科学进展, 2014, 35 (6): 45-52.

[8] 文华,高文,罗莉,等. 草鱼幼鱼的饲料苏氨酸需要量 [J]. 中国水产科学, 2009, 16 (2): 238-247.

[9] 胡晓霞. 生长中期草鱼的苏氨酸需要量研究 [D]. 雅安:四川农业大学, 2012.

[10] 冯琳. 苏氨酸对幼建鲤消化吸收能力和抗病力以及组织器官中蛋白质调控信号分子 TOR 表达的影响 [D]. 雅安:四川农业大学, 2010.

[11] Silva L C R, Furuya W M, Santos L D. Threonine levels in diets for Nile Tilapia [J]. Revista

[12] 周兴华, 向枭, 罗莉, 等. 吉富罗非鱼对饲料中苏氨酸的需要量 [J]. 淡水渔业, 2014, 44 (4): 83-89.

[13] Helland S J, Grisdale-Helland B G. Dietary threonine requirement of Atlantic salmon smolts [J]. Aquaculture, 2011, 321 (3-4): 230-236.

[14] Millamena O M, Bautista M N, Reyes O S, et al. Threonine requirement of juvenile marine shrimp *Penaeus monodon* [J]. Aquaculture, 2011, 151 (1): 9-14.

[15] Huai M Y, Tian L X, Liu Y J, et al. Quantitative dietary threonine requirement of juvenile Pacific white shrimp, *Litopenaeus vannamei* (Boone) reared in low-salinity water [J]. Aquaculture Research, 2009, 40 (8): 904-914.

[16] 王用黎. 凡纳滨对虾幼虾对苏氨酸、亮氨酸、色氨酸和缬氨酸需要量的研究 [D]. 湛江: 广东海洋大学, 2013.

[17] 王伟, 叶金云, 杨霞, 等. 中华绒螯蟹幼蟹对苏氨酸需要量的研究 [J]. 动物营养学报, 2015, 27 (2): 476-484.

[18] 李智, 徐博成, 汪以真. Meta 分析在动物生产性能评估中的应用 [J]. 动物营养学报, 2020, 32 (3): 1003-1009.

[19] Mercer L P. The quantitative nutrient-response relationship [J]. Journal of Nutrition, 1982, 112: 560-566.

[20] 王亚玲, 王常安, 刘红柏, 等. 苏氨酸的鱼类营养生理研究进展 [J]. 水产学杂志, 2021, 34 (4): 99-106.

[21] 王亚玲, 王常安, 刘红柏, 等. 低鱼粉饲料中添加苏氨酸对三倍体虹鳟生长性能、体成分及肌肉氨基酸组成的影响 [J]. 动物营养学报, 2021, 33 (4): 2390-2400.

[22] 高文. 草鱼幼鱼苏氨酸需要量的研究 [D]. 重庆: 西南大学, 2009.

[23] Aussansuwannakul A, Slider S D, Salem M, et al. Comparison of variable-blade to Allo-Kramer shear method in assessing rainbow trout (*Oncorhynchus mykiss*) fillet firmness [J]. Journal of Food Science, 2012, 77: 335-341.

[24] 白稚子, 刘明宇, 李树红, 等. 饲料中苏氨酸含量对建鲤肉质及组织蛋白酶 B、L 的影响 [J]. 食品与发酵工业, 2019, 45 (22): 90-96.

[25] Swatson H K, Gous R, Iji P A, et al. Effect of dietary protein level, amino acid balance and feeding level on growth, gastrointestinal tract, and mucosal structure of the small intestine in broiler chickens [J]. Animal Research, 2002, 51 (6): 501-515.

[26] 赵振山, 林可椒, 张益明, 等. 用 RNA/DNA 比值评定鲤鱼的生长及其配合饲料的营养价值 [J]. 水产学报, 1994, 18 (4): 257-264.

[27] Patra B C, Patra S, Bhattacharya M. Evaluating the nutritional condition of an indian climbing perch, *Anabas testudineus* fingerlings by the RNA/DNA, Ca/P Ratio and protein bio-synthesis in liver and muscle [J]. Fisheries and Aquaculture Journal, 2017, 8 (1): 811-815.

[28] 辛芳, 王雷, 刘梅, 等. 水产动物雷帕霉素受体信号通路的研究进展 [J]. 海洋科学, 2016, 40 (1): 147-154.

[29] Zou T, Cao S P, Xu W J, et al. Effects of dietary leucine levels on growth, tissue protein content and relative expression of genes related to protein synthesis in juvenile gibel carp (*Carassius auratus gibelio* var. CAS Ⅲ) [J]. Aquaculture Research, 2018, 49 (6): 2240-2248.

[30] Zhao J, Liu Y, Jiang J, et al. Effects of dietary isoleucine on the immune response antioxidant status

and gene expression in the head kidney of juvenile jian carp [J]. Fish & Shellfish Immunology, 2013, 35 (2): 572-580.

[31] Luo J B, Feng L, Jiang W D, et al. The impaired intestinal mucosal immune system by valine deficiency for young grass carp (*Ctenopharyngodon idella*) is associated with decreasing immune status and regulating tight junction proteins transcript abundance in the intestine [J]. Fish & Shellfish Immunology, 2014, 40 (1): 197-207.

[32] Feng L, Peng Y, Wu P, et al. Threonine affects intestinal function, protein synthesis and gene expression of TOR in Jian carp (*Cyprinus carpio* var. Jian) [J]. PLoS One, 2013, 8 (7): e69974

[33] 张凯凯. 大菱鲆幼鱼几种功能性氨基酸营养代谢的研究 [D]. 青岛: 中国海洋大学, 2014.

[34] 李雪吟. 赖氨酸对草鱼肠道免疫和结构屏障作用及其机制研究 [D]. 雅安: 四川农业大学, 2017.

[35] 洪杨. 苏氨酸对生长后期草鱼消化吸收功能和抗氧化能力影响的研究 [D]. 雅安: 四川农业大学, 2012.

[36] 彭艳. 晶体和包被处理的苏氨酸对幼建鲤生产性能、消化吸收功能和免疫功能影响的比较研究 [D]. 雅安: 四川农业大学, 2009.

[37] 黄瑾, 熊邦喜, 陈洁, 等. 鱼类消化酶活性及其影响因素的研究进展 [J]. 湖南农业科学, 2011, 5: 129-131, 141.

[38] 赵春蓉. 赖氨酸对幼建鲤消化能力和免疫功能的影响 [D]. 雅安: 四川农业大学, 2005.

[39] 帅柯. 蛋氨酸对幼建鲤消化能力和免疫功能的影响 [D]. 雅安: 四川农业大学, 2006.

[40] Zhao J, Liu Y, Jiang J, et al. Effects of dietary isoleucine on growth, the digestion and absorption capacity and gene expression in hepatopancreas and intestine of juvenile Jian carp (*Cyprinus carpio* var. Jian) [J]. Aquaculture, 2012, 368-369: 117-128.

[41] Habte-Tsion H M, Ren M C, Liu B, et al. Threonine influences the absorption capacity and brush-border enzyme gene expression in the intestine of juvenile blunt snout bream (*Megalobrama amblycephala*) [J]. Aquaculture, 2015, 448: 436-444.

[42] Kohen R, Nyska A. Invited review: Oxidation of biological systems: oxidative stress phenomena, antioxidants, redox reactions, and methods for their quantification [J]. Toxicologic pathology, 2002, 30 (6): 620-650.

[43] Habte-Tsion H M, Ren M, Liu B, et al. Threonine modulates immune response, antioxidant status and gene expressions of antioxidant enzymes and antioxidant-immune-cytokine-related signaling molecules in juvenile blunt snout bream (*Megalobrama amblycephala*) [J]. Fish & Shellfish Immunology, 2016, 51: 189-199.

[44] Liang H L, Mokrani A, Ji K, et al. Dietary leucine modulates growth performance, Nrf2 antioxidant signaling pathway and immune response of juvenile blunt snout bream (*Megalobrama amblycephala*) [J]. Fish & Shellfish Immunology, 2018, 73: 57-65.

[45] Ji K, Liang H L, Ren M C, et al. The immunoreaction and antioxidant capacity of juvenile blunt snout bream (*Megalobrama amblycephala*) involves the PI3K/Akt/Nrf2 and NF-κB signal pathways in response to dietary methionine levels [J]. Fish & Shellfish Immunology, 2020, 105: 126-134.

[46] Anderson D P. Environmental Factors in Fish Health: Immunological Aspects [J]. Fish Physiology, 1996, 15: 289-310.

[47] 封福鲜. 精氨酸、赖氨酸和苏氨酸对瓦氏黄颡鱼幼鱼生长、代谢及免疫力的影响 [D]. 青岛: 中国海洋大学, 2011.

[48] Carroll M C. The complement system in regulation of adaptive immunity [J]. Nature Immunology, 2004, 5: 981.

[49] Dong Y W, Feng L, Jiang W D, et al. Dietary threonine deficiency depressed the disease resistance, immune and physical barriers in the gills of juvenile grass carp (*Ctenopharyngodon idella*) under infection of *Flavobacterium columnare* [J]. Fish & Shellfish Immunology, 2018, 72: 161-173.

[50] Habte-Tsion H M, Ge X P, Liu B, et al. A deficiency or an excess of dietary threonine level affects weight gain, enzyme activity, immune response and immune-related gene expression in juvenile blunt snout bream (*Megalobrama amblycephala*) [J]. Fish & Shellfish Immunology, 2015, 42 (2): 439-446.

[51] Kato G, Takano T, Sakai T, et al. Vibrio anguillarum bacterin uptake via the gills of Japanese flounder and subsequent immune responses [J]. Fish & Shellfish Immunology, 2013, 35 (5): 1591-1597.

第六章 精氨酸

精氨酸是含有两个碱性基团及氨基和胍基的脂肪族氨基酸，是水产动物体内十种必需氨基酸之一。1895 年，Hedin 首次在哺乳动物的蛋白质中发现了精氨酸的存在[1]。精氨酸在自然界中有两种异构体存在：D-精氨酸和 L-精氨酸是动物机体内精氨酸存在的两种异构体形式，前者在动物机体内发挥着营养、代谢和免疫等主要生理功能。水产动物内源性精氨酸合成能力相对较弱，因此饲料中需添加一定水平的精氨酸来维持机体正常供应[2]。精氨酸缺乏通常会减缓水产动物生长速度、降低饲料转化率及减少蛋白质沉积，鱼类精氨酸需要量在 1.6%～7.34%（占饲料）变化[3]，主要是由于氨基酸类型、吸收率以及饲料能量的不同都会影响氨基酸的需要量[4,5]。此外，精氨酸在调节免疫功能方面具有非常重要的作用[6]。本章综述了近年来精氨酸对鱼类生长、体成分及免疫功能的影响以及精氨酸和赖氨酸的平衡，旨在为精氨酸在鱼类生产中的应用提供参考。

第一节 精氨酸的理化特性及在体内的代谢途径

精氨酸是一种碱性氨基酸，分子式为 $C_6H_{14}N_4O_2$，化分子质量为 174.20，化学名为 2-氨基-5-胍基戊酸，为白色晶体或晶体状粉末，天然精氨酸主要以 L-精氨酸形式存在。在水中结晶的产物含 2 分子的结晶水，105 ℃失去结晶水。在乙醇中结晶的产物是无水物，238 ℃分解，溶于水，微溶于乙醇，不溶于乙醚。机体内精氨酸主要来源是饲料、机体蛋白质的分解以及机体内其他氨基酸（谷氨酸、瓜氨酸等）的合成。

精氨酸是机体内唯一含有胍基的氨基酸，代谢过程中可以生成肌酸，同时存储高能量磷酸盐使肌肉三磷酸腺苷（ATP）再生[7]。精氨酸是合成一氧化氮（NO）的前体物质，在内皮型一氧化氮合酶（eNOS）和诱导型一氧化氮合酶（iNOS）的作用下生成 NO 和瓜氨酸。饲料中添加精氨酸显著提高鱼血清中总一氧化氮合酶（T-NOS）的活性和肝脏中 T-NOS 和 iNOS 的活性，进而生成更多的 NO[4]。NO 在系统水平和细胞水平都是一种重要的多效信号分子，参与机体的免疫、血管生成及基因表达。精氨酸参与氮代谢，在精氨酸酶（尿素循环中一种重要的酶）的作用下生产尿素氮和鸟氨酸（Ornithine，Orn），研究表明，精氨酸水平的增加可以显著提高肝脏中精氨酸酶的活性[7,8,9]。Berge 等[10]研究表明，高精氨酸组肌肉和血清中 Orn 的含量也高于其他组，且肌肉中精氨酸和 Orn 呈正相关（$Orn=-0.0013+0.1484\times$精氨酸，$r=0.925$）。Orn 在鸟氨酸脱羧酶作用下生产多胺，也可以在吡咯啉-5-羧酸还原酶和吡咯啉-5-羧酸脱氨酶的作用下生产脯氨酸和谷氨酰胺。饲料中添加精氨酸显著提高血清中谷氨酰胺的含量，脯氨酸的含量也有提高的趋势[11]。Pohlenz 等[12]研究表明，随着饲料精氨酸添加量的增加，血清中 Orn、脯氨酸、瓜氨酸、谷氨酸及谷氨酰胺的含量也都显著提高。精氨酸代谢如图 6-1[13]。

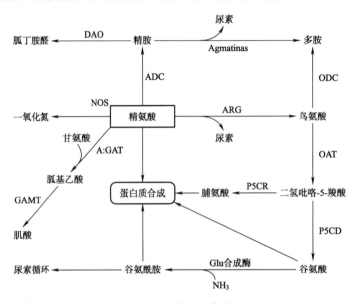

图 6-1 精氨酸代谢

注：ADC 表示精氨酸脱羧酶，A:GAT 表示精氨酸-甘氨酸转脒酶，ARG 表示精氨酸酶，NOS 表示一氧化氮合成酶，OAT 表示鸟氨酸转氢酶，P5CR 表示二氢吡咯-5-羧酸还原酶，ODC 表示鸟氨酸脱羧酶，P5CD 表示二氢吡咯-5-羧酸脱氢酶，GAMT 表示 S-腺苷蛋氨酸-胍基乙酸 N-甲基转移酶，DOA 表示二胺氧化酶。

第二节　精氨酸需要量及其对生长性能的影响

表 6-1 总结了近年来关于水产动物精氨酸需要量的研究报道。对于大多数鱼类，精氨酸需要量在 1.51%～3.39%（占饲料）范围内；对于甲壳类动物，精氨酸需要量在 1.66%～3.62%（占蛋白）范围内。

表 6-1 水产动物蛋氨酸需要量

品种	体重（g）	模型	指标	需要量（占饲料干物质，%）	参考文献
鱼类					
杂交鲟	22.02	折线模型	增重率、特定生长率、蛋白质沉积率	2.91～2.98	李培佳等[14]
大菱鲆	43.07	折线模型	特定生长率	3.17	付丰顺等[15]
许氏平鲉	12.03	二次多项式模型	增重率	2.78	沈钰博等[16]
驼背鲈	6.55	二次多项式模型	增重率	3.39	穆伟等[17]
团头鲂	20.0	二次多项式模型	饲料系数、特定生长率	1.85	梁化亮[18]
黄颡鱼	1.13	二次多项式模型	增重率	2.74	赵红霞等[19]
			一氧化氮含量	2.94	

(续)

品种	体重（g）	模型	指标	需要量（占饲料干物质，%）	参考文献
点带石斑鱼	7.5	二次多项式模型	增重率	3.05	Han 等[20]
杂交鲟	3.6	折线模型	特定生长率、饲料效率	2.48	Wang 等[21]
吉富罗非鱼	81.52	二次多项式模型	增重率、饲料系数、蛋白质效率	1.51~1.58	武文一等[22]
虾蟹类					
凡纳滨对虾	0.50	二次多项式模型	增重率	2.16	曾雯娉[23]
罗氏沼虾	1.55	二次多项式模型	增重率	2.94	吴骏[24]
克氏原螯虾	7.60	二次多项式模型	增重率	1.66	张微微[25]
中华绒螯蟹	2.03	二次多项式模型	增重率	3.62	叶金云等[26]

饲料中添加精氨酸促进鱼的生长，一方面是由于精氨酸能够改善鱼肠道的形态结构，研究表明，1%精氨酸能提高肠道末端绒毛高度和肠上皮细胞高度，2%精氨酸提高肠道中段和末端褶皱高度[27]；另一方面是由于精氨酸能提高胃、前肠及中肠胃蛋白酶和胰蛋白酶的活性，同时可以提高胃和前肠中淀粉酶的活性，进而提高饲料的消化和吸收率[28]。此外，精氨酸还可以促进生长激素和胰岛素的分泌，进而促进鱼类的生长[29]。

饲料中添加1%精氨酸显著提高杂交条纹鲈末重、特定生长率及饲料转化率，成活率具有提高的趋势[30]。印度鲇饲料中精氨酸含量在1.60%时，增重率、饲料效率及蛋白质效率都最高，以增重率、饲料效率、蛋白质效率及鱼体蛋白质含量分别进行二元回归分析，精氨酸的需要量分别是1.68%、1.63%、1.61%和1.61%，综合上述结果得出印度鲇饲料精氨酸最适含量是1.63%[31]。卡特拉鲃饲料中精氨酸含量在1.75%时增重率、蛋白质沉积率、饲料系数及精氨酸沉积量显著高于其他组[32]。Khan[33]同时研究表明印度囊鳃鲇饲料精氨酸含量在1.75%时，饲料效率和蛋白质沉积率都最高。Ren 等[34]以特定生长率和饲料效率进行二元回归分析，表明军曹鱼最适精氨酸含量为2.85%和2.82%。相反，也有一些研究表明精氨酸对鱼类的生长性能没有影响或有负面影响[35]，其原因可能主要有两方面：一方面由于试验鱼的品种不同，另一方面由于试验持续时间不同。

RNA/DNA 比值是反映鱼类生长的一个可靠的指标。动物细胞 DNA 的含量相对恒定，但 RNA 的含量与机体蛋白质合成率密切相关[36]。蛋白质的合成量与 RNA 的含量变化一致，因此很多鱼类试验用 RNA/DNA 的比值来反映生长[37,38]。Zehra 和 Khan[32]研究表明，精氨酸添加量在1%~1.75%时，肌肉 DNA 含量线性降低，而 RNA 含量以及 RNA/DNA 比值提高，这与 Abidi[38]的研究结果一致。

第三节　精氨酸对鱼类体成分的影响

一、精氨酸对鱼体粗蛋白、粗脂肪含量的影响

精氨酸含量低于最适量时，鱼体粗蛋白含量随着饲料精氨酸的增加而提高，但精氨酸含量高于最适量时，鱼体粗蛋白含量随着精氨酸添加量的增加而降低。Zhou 等[4]研究表

明，饲料精氨酸含量在 2.01%～3.08%时，肌肉中粗蛋白的含量随着精氨酸的增加而提高；但精氨酸含量在 3.08%～3.27%时，随着精氨酸含量的增加而降低。Zehra 和 Khan[32]研究表明，鱼体蛋白质含量与精氨酸添加量呈二次曲线关系，精氨酸含量在 1.75%时蛋白质沉积量最高。Ren 等[34]研究表明，饲料中精氨酸含量在 1.76%时，鱼体粗蛋白含量最低；2.96%时，鱼体蛋白质含量最高；高于 2.96%时，粗蛋白含量又降低。Luo 等[39]研究表明，精氨酸对全鱼粗蛋白含量的影响没有达到显著水平，但肌肉和肝脏粗蛋白含量显著提高。精氨酸提高蛋白质含量的原因可能是由于精氨酸及代谢产生的谷氨酰胺和 NO 可以激活肌肉中哺乳动物雷帕霉素靶蛋白（mTOR）信号途径，mTOR 激活后会促进磷酸化核糖体 S6 蛋白激酶（p70s6 激酶）和真核生物启动子 4E－结合蛋白1（eIF4E－BP1）磷酸化，然后形成用于多肽合成的激活启动复合物，进而促进蛋白质合成[39,40]。

鱼体粗脂肪含量随着饲料精氨酸添加量的增加而降低。Zehra 和 Khan[32]研究表明，粗脂肪含量与精氨酸添加量呈线性关系，随着精氨酸添加量的增加（1%～2.25%），粗脂肪含量线性降低（4.68%～2.84%），然而对粗灰分的含量没有影响。Luo 等[39]研究表明，饲料添加精氨酸显著降低鱼体粗脂肪含量，且肝脏水分含量显著降低。精氨酸通过内分泌信号调节腺苷酸活化蛋白激酶（AMPK）来抑制脂肪酸和类固醇的合成，进而降低粗脂肪的含量[29]。

二、精氨酸对鱼体氨基酸组成的影响

Zhou 等[28]研究表明，饲料中添加精氨酸提高黑鲷肌肉中亮氨酸、异亮氨酸和精氨酸的含量，对其他必需氨基酸的含量没有显著影响，总必需氨基酸的含量也显著提高；非必需氨基酸中谷氨酸、丙氨酸及丝氨酸的含量显著提高，对非必需氨基酸总含量的影响没有达到显著水平。Luo 等[39]研究表明，精氨酸显著提高全鱼必需氨基酸、非必需氨基酸及总氨基酸的含量，氨基酸沉积率也显著提高（除缬氨酸和异亮氨酸外）。Zhou 等[4]研究表明，精氨酸对石斑鱼肝脏必需氨基酸组成没有显著影响，饲料精氨酸含量 3.08%时，肌肉必需氨基酸和精氨酸的含量显著高于 2.01%组。

第四节 精氨酸对鱼类免疫功能的影响

一、精氨酸对免疫系统的促进作用

精氨酸作为 NO 的唯一前体物质，在鱼类非特异免疫系统和特异免疫系统起着非常重要的作用。

高精氨酸能提高天然杀伤细胞毒性及白介素-2 的含量，同时提高 T-细胞 CD3 的蛋白表达[40,41]。此外，精氨酸还能够提高巨噬细胞和中性粒细胞的吞噬作用和杀伤能力，调节淋巴细胞亚群的黏附分子、趋药性及细胞增殖[42]。Cheng 等[30]研究表明，精氨酸显著提高杂交条纹鲈细胞外超氧阴离子和中性粒细胞氧自由基的产生，细胞内超氧阴离子的含量也有提高的趋势，但差异没有达到显著水平。精氨酸缺乏导致肾脏巨噬细胞产生的超氧离子和血液中性粒细胞产生的氧自由基的量显著降低[12]。一些研究表明饲料精氨酸含量影响鱼类血液组成[43]，如血细胞压积、红细胞数及白细胞数，这些指标可以反映机体的造血功能和免疫功能[38]。饲料添加4%精氨酸显著提高血液中血红蛋白含量、红细胞比容及红细胞数[43]。Cheng等[30]研究表明，1%精氨酸能显著提高杂交条纹鲈血清溶菌酶的活性，溶菌酶是激活补体系统和吞噬细胞的免疫调理素。

Pohlenz等[44]研究表明，斑点叉尾鮰接种爱德华氏菌7 d后，添加4%精氨酸组血清爱德华氏菌抗体效价显著提高，且脾脏和头肾中蛋白质含量提高；14 d后，精氨酸组脾脏和头肾淋巴细胞抗爱德华氏菌功能提高。Buentello和Gatlin[45]研究表明，斑点叉尾鮰感染爱德华氏菌后，2%精氨酸饲料能显著提高其成活率。

二、精氨酸的抗氧化作用

水产动物精氨酸缺乏会导致机体抗氧化能力降低。对团头鲂和鲤的研究发现，饲喂缺乏精氨酸引起生长性能和抗氧化酶活性下降，肠道丙二醛水平升高[46,47]。同高等动物类似，精氨酸主要是通过激活Nrf2信号通路来调控水产动物机体抗氧化能力[48]。在对建鲤的研究中发现，适度增加饲料精氨酸水平可以激活Nrf2信号通路，进而调控下游的抗氧化酶或非酶物质的活力/含量及基因表达从而增加草鱼机体抵抗氧化应激的能力[49]。此外，在氧化应激条件下（毒物暴露、氧化油使用等），精氨酸抗氧化作用的潜力能够被更好地反映[50]。对草鱼的研究发现，在精氨酸缺乏时，铜会诱导氧化损伤导致鱼鳃和鳃细胞凋亡，但补充精氨酸后对铜诱导鳃损伤具有预防作用，同时会激活Nrf2信号通路，提高抗氧化基因（$SOD1$、CAT、GPx和GST等）的基因表达[51]。

三、精氨酸对炎症反应的影响

众多研究发现精氨酸在转录水平上调节水产动物的炎症反应。一些研究表明无论在正常或应激条件下，精氨酸水平的不足或过剩都会导致水产动物促炎症因子基因（IL-1β、TNF-α和IL-8）的表达下降[52]。对建鲤的研究发现，促炎细胞因子的基因表达随饲料精氨酸水平的增加呈剂量依赖性，而抗炎细胞因子基因的表达只有高水平的精氨酸条件下显著增加，这可能与抵消急性炎症负面影响的反馈机制有关[53]。相反，饲料中添加精氨酸可下调松浦镜鲤肠促炎细胞因子基因表达（TNF-α、IL-1β和IL-8），上调抗炎细胞因子基因（TGF-β和IL-10）[47]。此外，对大西洋鲑[54,55]、建鲤[56]的研究发现，精氨酸的抗炎特性能够抑制脂多糖引发的组织炎症。

四、精氨酸对水体氨毒性的缓解作用

精氨酸参与机体内尿素循环和氨转化为尿素的过程，因此饲料精氨酸水平能够成为缓解水体氨毒性的关键因素。Chen等[57]研究发现与5.5%（占饲料蛋白）精氨酸对照组相比，黄颡鱼饲料中精氨酸水平为6.6%（占饲料蛋白）时可降低氨胁迫72 h后死亡率。虽然关于精氨酸缓解水体氨毒性的具体机制目前研究还未见报道，但可能存在两种不同的潜在机理。首先，在对猫[58]和雪貂[59]的研究中发现，适宜水平精氨酸的补充激活了尿素循环，从而增加了对氨的解毒作用。其次，在对大菱鲆和建鲤的研究中发现，适宜水平精氨酸的补充能够缓解氨毒性对机体造成的不良影响，如生理应激和氧化应激[60,61]。

第五节 赖氨酸与精氨酸、谷氨酰胺的交互作用

必需氨基酸包括精氨酸能提高鱼类的生长性能。然而,必需氨基酸的含量过高又将导致氨排放增加,从而破坏水质且影响生长[62]。因此,只有氨基酸平衡才能促进鱼类的生长。Kaushik 和 Fauconeau[63]研究表明,精氨酸和赖氨酸在肠道利用同一种转运载体进行转运,因此在吸收、转运及代谢方面存在竞争抑制现象,虹鳟血清精氨酸水平随着赖氨酸添加量的增加而降低。军曹鱼饲喂赖氨酸和精氨酸比例为 0.8、1.1 和 1.8 的高植物蛋白饲料,结果表明,赖氨酸和精氨酸比例为 1.1 组鱼的体重、体长及增重率都高于 0.8 和 1.8 组[64]。另有研究表明,饲料过量赖氨酸对斑点叉尾鮰、欧洲鲈及牙鲆的生长性能及血清精氨酸水平没有负面影响[65,66]。在实际生产中不同鱼种在配制饲料时须考虑精氨酸和赖氨酸的平衡,既有利于鱼类的生长,又可以节约饲料成本。

在肠细胞中,谷氨酰胺可能通过中间体吡咯啉-5-羧酸盐转化为瓜氨酸。瓜氨酸在肾脏中被转化为精氨酸,然后运输到肝脏,通过尿素循环参与机体代谢,或用于生产多胺和肌酸[67]。谷氨酸对饲料精氨酸的节约效应已在斑点叉尾鮰[68]的研究得到证实,在精氨酸缺乏(2%,占饲料蛋白)饲料中添加适宜水平谷氨酰胺(2.08%,占饲料蛋白)显著提高饲料效率及血浆瓜氨酸和精氨酸水平。在对鲤[47]的研究中同样发现,当氨酸水平为 3.8%(占饲料蛋白)时,饲料中补充 0.37%~0.62%(占饲料蛋白)谷氨酰胺能够更好地促进鱼体生长。这些研究表明,谷氨酰胺能够有效缓解精氨酸缺乏引起的抑制生长效应,然而这是否适用于所有鱼类还需要进一步研究。此外,瓜氨酸在鱼体内转为精氨酸的具体机制也需进一步深入研究。

第六节 小结与展望

综上所述,适量添加精氨酸能够提高鱼类生长性能,改善鱼体营养成分,提高机体免疫力。目前对水产动物精氨酸合成能力的研究相对较少,有研究认为谷氨酰胺可能是少数鱼类精氨酸合成的来源,但有待深入研究;精氨酸可以通过刺激氨转化为尿素来促进氨的解毒作用,但尿素循环在水产动物机体内精氨酸合成代谢中的具体机制仍不明确。此外,在水产动物精氨酸需要量研究中需考虑生长阶段、饲料营养组成、功能性营养物质(如谷氨酰胺和其他抗氧化剂)、饲养条件(如饲养密度、水流率)和一些外部因素(如压力、氨水平和疾病)。因此,研究精氨酸对不同水产动物的抗应激作用是必要的。为了阐明精氨酸对水产动物免疫系统的作用机制,需要进一步评估正常和致病条件下精氨酸对水产动物的免疫活性物质和转录组的影响。

参考文献

[1] 孙红暖,杨海明,王志跃,等. 精氨酸对动物的营养生理及免疫作用 [J]. 动物营养学报,2014,26 (1):54-62.

[2] Wilson R P. Amino acids and proteins [J]. Fish nutrition, 2002: 143.

[3] Fagbenro O A, Nwanna L C, Adebayo O T. Dietary arginine requirement of the African catfish, *Clarias gariepinus* [J]. Journal of Applied Aquaculture, 1999, 9: 59-64.

[4] Zhou Q C, Zeng W P, Wang H L, et al. Dietary arginine requirement of juvenile yellow grouper *Epinephelus awoara* [J]. Aquaculture, 2012, 350-353: 175-182.

[5] Simmons L, Moccia R D, Bureau D P. Dietary methionine requirement of juvenile Arctic charr *Salvelinus alpinus* (L.) [J] Aquaculture Nutrition, 1999, 5: 93-100.

[6] Jobgen W S, Fried S K, Fu W J, et al. Regulatory role for the arginine-nitric oxide pathway in metabolism of energy substrates [J]. The Journal of Nutritional Biochemistry, 2006, 17: 571-588.

[7] Denis E, Frcsi M D, Lieberman M D, et al. Immunonutrition: the role of arginine [J]. Nutrition, 1998, 14: 7-8.

[8] Berge G E, Lied E, Sveier H. Nutrition of Atlantic salmon (*Salmo salar*): the requirement and metabolism of arginine [J]. Comparative Biochemistry and Physiology part A, 1997, 117: 501-509.

[9] Tulli F, Vachot C, Tibaldi E, et al. Contribution of dietary arginine to nitrogen utilization and excretion in juvenile sea bass (*Dicentrarchus labrax*) fed diets differing in protein source [J]. Comparative Biochemistry and Physiology Part A, 2007, 147: 179-188.

[10] Berge G E, Sveier H, Lied E. Effects of feeding Atlantic salmon (*Salmo salar* L.) imbalanced levels of lysine and arginine [J]. Aquaculture Nutrition, 2002, 8: 239-248.

[11] Tesser M B, Terjesen B F, Zhang Y, et al. Free- and peptide-based dietary arginine supplementation for the South American fish pacu (*Piaractus mesopotamicus*) [J]. Aquaculture Nutrition, 2005, 11: 443-453.

[12] Pohlenz C, Buentello A, Helland S J, et al. Effects of dietary arginine supplementation on growth, protein optimization and innate immune response of channel catfish *Ictalurus punctatus* (Rafinesque 1818) [J]. Aquaculture Research, 2012, 45 (3): 491-500.

[13] Carmelo N J, Bobbi L H. Arginine and immunity: a unique perspective [J]. Biomed Pharmacother, 2002, 56: 471-482.

[14] 李培佳, 陈晓瑛, 赵红霞, 等. 精氨酸对杂交鳢生长性能、体组成、血浆生化指标及抗氧化能力的影响 [J]. 动物营养学报, 2022, 34 (3): 1820-1830.

[15] 付丰顺, 刘成栋, 王旋, 等. 大菱鲆对饲料精氨酸的需要量及饲料精氨酸水平对大菱鲆生长和代谢的影响 [J]. 水产学报, 2021, 45 (10): 1692-1702.

[16] 沈钰博, 王际英, 李宝山, 等. 许氏平鲉幼鱼对饲料中精氨酸需要量的研究 [J]. 渔业科学进展, 2022, 1-13.

[17] 穆伟. 驼背鲈幼鱼最适精氨酸、赖氨酸和蛋氨酸需要量研究 [D]. 海口: 海南大学, 2020.

[18] 梁化亮. 精氨酸对团头鲂幼鱼生长、营养代谢和免疫功能的影响及作用机制 [D]. 南京: 南京农业大学, 2019.

[19] 赵红霞, 陈启明, 黄燕华, 等. 饲料精氨酸水平对黄颡鱼幼鱼生长性能、消化吸收相关指标、免疫功能和抗氧化能力的影响 [J]. 动物营养学报, 2018, 30 (12): 5040-5051.

[20] Han F, Chi S, Tan B, et al. Metabolic and immune effects of orange-spotted grouper, *Epinephelus coioides* induced by dietary arginine [J]. Aquaculture Reports, 2018, 10: 8-16.

[21] Wang L S, Wu J, Wang C A, et al. Dietaryarginine requirement of juvenile hybrid sturgeon (*Acipenser schrenckii*×*Acipenser baerii*). Aquaculture Research [J], 2017, 48: 5193-5201.

[22] 武文一, 蒋明, 刘伟, 等. 吉富罗非鱼对饲料精氨酸的需要量 [J]. 动物营养学报, 2016, 28 (5):

1412‑1424.

[23] 曾雯娉. 凡纳滨对虾幼虾对赖氨酸、蛋氨酸、精氨酸和苯丙氨酸需要量的研究 [D]. 湛江: 广东海洋大学, 2012.

[24] 吴骏. 罗氏沼虾幼虾精氨酸适宜需要量及饲料精氨酸/赖氨酸不同配比试验研究 [D]. 扬州: 扬州大学, 2016.

[25] 张微微. 克氏原螯虾对赖氨酸和精氨酸需要量的研究 [D]. 南京: 南京农业大学, 2012.

[26] 叶金云, 王友慧, 郭建林, 等. 中华绒螯蟹对赖氨酸、蛋氨酸和精氨酸的需要量 [J]. 水产学报, 2010, 34 (10): 1541‑1548.

[27] 李晋南, 张圆圆, 范泽, 等. 饲料精氨酸水平对松浦镜鲤幼鱼生长、抗氧化能力和肠道消化酶活性及其组织学结构的影响 [J]. 水产学杂志, 2021, 34 (5): 32‑39.

[28] Zhou F, Zhou J, Shao Q J, et al. Effects of Arginine‑deficient and replete diets on growth performance, digestive enzyme activities and genes expression of black sea bream, *Acanthopagrus schlegelii*, Juveniles [J]. Journal of The World Aquaculture Society, 2012, 43: 828‑839.

[29] Andoh T. Stress inhibits insulin release induced by feeding and arginine injection in barfin flounder *Verasper moseri* [J]. Fisheries Science, 2014, 80: 311‑316.

[30] Cheng Z, Gatlin Ⅲ D M, Buentello A. Dietary supplementation of arginine and/or glutamine influences growth performance, immune responses and intestinal morphology of hybrid striped bass (*Morone chrysops*×*Morone saxatilis*) [J]. Aquaculture, 2012, 362‑363: 39‑43.

[31] Ahemd I. Dietary arginine requirement of fingerling Indian catfish (*Heteropneustes fossilis*, Bloch) [J]. Aquaculture International, 2013, 21: 255‑271.

[32] Zehra S, Khan M A. Dietary Arginine Requirement of fingerling Indian Major Carp, *Catla catla* (Hamilton) [J]. Journal of The World Aquaculture Society, 2013, 44: 363‑373.

[33] Khan M A. Effects of dietary arginine levels on growth, feed conversion, protein productive value and carcass composition of stinging catfish fingerling *Heteropneustes fossilis* (Bloch) [J]. Aquaculture International, 2012, 20: 935‑950.

[34] Ren M C, Ai Q H, Mai K S. Dietary arginine requirement of juvenile cobia (*Rachycentron canadum*) [J]. Aquaculture Research, 2012, 45 (2): 225‑233.

[35] Cheng Z Y, Buentello J A, Gatlin Ⅲ D M. Effects of dietary arginine and glutamine on growth performance, immune responses and intestinal structure of red drum, *Sciaenops ocellatus* [J]. Aquaculture, 2011, 319: 247‑252.

[36] Tanaka Y, Gwak W S, Tanaka M, et al. Ontogenetic changes in RNA, DNA and protein contents of laboratory‑reared Pacific bluefin tuna *Thunnus orientalis* [J]. Fisheries Science, 2007, 73: 378‑384.

[37] Mercaldo‑Allen R, Kuropat C, Caldarone E M. An RNA: DNA‑based growth model for young‑of‑the‑year winter flounder *Pseudopleuronectes americanus* (Walbaum) [J]. Journal of Fish Biology, 2008, 72: 1321‑1331.

[38] Abidi S F, Khan M A. Dietary arginine requirement of fingerling Indian major carp, *Labeo rohita* (Hamilton) based on growth, nutrient retention efficiencies, RNA/DNA ratio and body composition [J]. Journal of Applied Ichthyology, 2009, 25: 707‑714.

[39] Luo Z, Liu Y J, Mai K S, et al. Effects of dietary arginine levels on growth performance and body composition of juvenile grouper *Epinephelus coioides* [J]. Journal of Applied Ichthyology, 2007, 23: 252‑257.

[40] Li P, Yin Y L, Li D, et al. Amino acids and immune function [J]. British Journal of Nutrition, 2007, 98: 237-252.

[41] Choi B S, Martinez-Falero I C, Corest C, et al. Differential impact of l-arginine deprivation on the activation and effector functions of T cells and macrophages [J]. Journal of Leukocyte Biology, 2009, 85: 268-277.

[42] Abdukalykova S T, Zhao X, Ruiz-Feria C A. Arginine and vitamin E modulate the subpopulations of T lymphocytes in broiler chickens [J]. Poultry Science, 2008, 87: 50-55.

[43] Zhou F, Shao Q J, Xiao J X, et al. Effects of dietary arginine and lysine levels on growth performance, nutrient utilization and tissue biochemical profile of black sea bream, *Acanthopagrus schlegelii*, fingerlings [J]. Aquaculture, 2010, 319: 72-80.

[44] Pohlenz C, Buentello A, Crisctiello R M F, et al. Synergies between vaccination and dietary arginine and glutamine supplementation improve the immune response of channel catfish against *Edwardsiella ictaluri* [J]. Fish & Shellfish Immunology, 2012, 33: 543-551.

[45] Buentello U J A, Gatlin III D M. Effects of elevated dietary arginine on resistance of channel catfish to exposure to *Edwardsiella ictaluri* [J]. Journal of Aquatic Animal Health, 2001, 13: 194-201.

[46] Ren M, Liao Y, Xie J, et al. Dietary arginine requirement of juvenile blunt snout bream, *Megalobrama amblycephala* [J]. Aquaculture, 2013, 414: 229-234.

[47] Wang L S, Li J N, Wang C A, et al. Effect of N-carbamoylglutamate supplementation on the growth performance, antioxidant status and immune response of mirror carp (*Cyprinus carpio*) fed an arginine-deficient diet [J]. Fish & Shellfish Immunology, 2019, 84: 280-289.

[48] Wang B, Liu Y, Feng L, et al. Effects of dietary arginine supplementation on growth performance, flesh quality, muscle antioxidant capacity and antioxidant-related signalling molecule expression in young grass carp (*Ctenopharyngodon idella*) [J]. Food Chemistry, 2015, 167: 91-99.

[49] Wang B, Feng L, Chen G F, et al. Jian carp (*Cyprinus carpio* var. Jian) intestinal immune responses, antioxidant status and tight junction protein mRNA expression are modulated via Nrf2 and PKC in response to dietary argine deficiency [J]. Fish & shellfish Immunology, 2016, 51: 116-124.

[50] Lin C C, Tsai W C, Chen J Y, et al. Supplements of L-arginine attenuate the effects of high-fat meal on endothelial function and oxidative stress [J]. International Journal of Cardiology, 2008, 127: 337-341.

[51] Wang B, Feng L, Jiang W D, et al. Copper-induced tight junction mRNA expression changes, apoptosis and antioxidant responses via NF-xB, TOR and NrF2 signaling molecules in the gills of fish: preventive role of arginine [J]. Aquatic Toxicology, 2015, 158: 125-137.

[52] Hoseini S M, Yousefi M, Hoseinifar S H, et al. Effects of dietary arginine supplementation on growth, biochemical, and immunological responses of common carp (*Cyprinus carpio* L.), stressed by stocking density [J]. Aquaculture, 2019, 503: 452-459.

[53] Chen G, Liu Y, Jiang J, et al. Effect of dietary arginine on the immune response and gene expression in head kidney and spleen following infection of Jian carp with *Aeromonas hydrophila* [J]. Fish & Shellfish Immunology, 2015, 44: 195-202.

[54] Holen E, Espe M, Andersen S M, et al. A coculture approach show that polyamine turnover is affected during inflammation in Atlantic salmon immune and liver cells and that arginine and LPS exerts opposite effects on p38MAPK signaling [J]. Fish & Shellfish Immunology, 2014, 37: 286-298.

[55] Martins G P, Espe M, Zhang Z, et al. Surplus arginine reduced lipopolysaccharide induced

transcription of proinflammatory genes in Atlantic salmon head kidney cells [J]. Fish & Shellfish Immunology, 2019, 86: 1130-1138.

[56] Jiang J, Shi D, Zhou X-Q, et al. *In vitro* and *in vivo* protective effect of arginine against lipopolysaccharide induced inflammatory response in the intestine of juvenile Jian carp (*Cyprinus carpio* var. Jian) [J]. Fish & Shellfish Immunology, 2015, 42: 457-464.

[57] Chen Q, Zhao H, Huang Y, et al. Effects of dietary arginine levels on growth performance, body composition, serum biochemical indices and resistance ability against ammonia-nitrogen stress in juvenile yellow catfish (*Pelteobagrus fulvidraco*) [J]. Animal nutrition, 2016, 2: 204-210.

[58] Morris J G, Rogers Q R. Ammonia intoxication in the near-adult cat as a result of a dietary deficiency of arginine [J]. Science, 1978, 199: 431-432.

[59] Deshmukh D R, Shope T C. Arginine requirement and ammonia toxicity in ferrets [J]. The Journal of Nutrition, 1983, 113: 1664-1667.

[60] Costas B, Rego P, Conceicao L, et al. Dietary arginine supplementation decreases plasma cortisol levels and modulates immune mechanisms in chronically stressed turbot (*Scophthalmus maximus*) [J]. Aquaculture Nutrition, 2013, 19: 25-38.

[61] Hart S D, Brown B J, Gould N L, et al. Predicting the optimal dietary essential amino acid profile for growth of juvenile yellow perch with whole body amino acid concentrations [J]. Aquaculture Nutrition, 2010, 16: 248-253.

[62] Kaushik S J, Fauconeau B. Effects of lysine administration on plasma arginine and on some nitrogenous catabolites in rainbow trout [J]. Comparative Biochemistry and Physiology part A, 1984, 79: 459-462.

[63] Nguyen M V, Rønnestad IBL, Lai H V, et al. Imbalanced lysine to arginine ratios reduced performance in juvenile cobia (*Rachycentron canadum*) fed high pant protein diets [J]. Aquacultre Nutrition, 2014, 20: 25-35.

[64] Robinson E H, Wilson R P, Poe W E. Arginine requirement and apparent absence of a lysine-arginine antagonist in fingerling channel catfish [J]. The Journal of Nutrition, 1981, 111: 46-52.

[65] TibaldiI E, Tulli F, Lanari D. Arginine requirement and effect of different dietary arginine and lysine levels for fingerling sea bass (*Dicentrarchus labrax*) [J]. Aquaculture, 1994, 127: 207-218.

[66] Wu G, Bazer F W, Davis T A, et al. Arginine metabolism and nutrition in growth, health and disease [J]. Amino Acids, 2009, 37: 153-168.

[67] Buentello JA, Gatlin DM Ⅲ. The dietary arginine requirement of channel catfish (*Ictalurus punctatus*) is influenced by endogenous synthesis of arginine from glutamic acid [J]. Aquaculture, 2000, 188: 311-321.

第七章 亮氨酸

必需氨基酸在水产动物生长和健康方面发挥重要作用。支链氨基酸（Branched-chain amino acid，BCAA）属于必需氨基酸。由亮氨酸、缬氨酸和异亮氨酸组成的支链氨基酸，在动物蛋白质总氨基酸中的占比为18%～20%，主要在骨骼肌中被支链氨基酸脱氢酶复合物（BCAA dehydrogenase enzyme complex）氧化，为肌肉内源合成谷氨酰胺提供α-氨基[1,2]。亮氨酸在机体合成蛋白质、能量代谢、葡萄糖平衡等方面具有重要作用[3]，是一种功能性氨基酸。亮氨酸通过激活雷帕霉素靶蛋白（TOR）信号通路，调控机体蛋白质合成、分解代谢和免疫功能[4]。同时，亮氨酸在血红蛋白生成及胁迫条件下调节血糖水平方面具有重要作用[5]。本章综述了水产动物亮氨酸的需要量、亮氨酸对蛋白质代谢、抗氧化、免疫功能、肠道发育的影响，以期为亮氨酸在水产动物营养需要、功能机理与健康养殖方面的深入研究提供参考。

第一节 亮氨酸的理化性质

亮氨酸（Leucine，Leu），分子式$C_6H_{13}NO_2$，分子质量131.18，分子结构如图7-1所示，熔点大于300℃，白色结晶或结晶性粉末，无臭，味微苦。易溶于甲酸，略溶于水，极微溶于乙醇或乙醚。

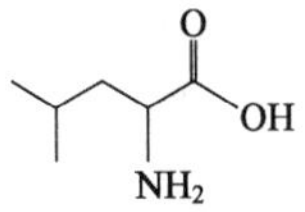

图7-1 亮氨酸分子结构

第二节 水产动物亮氨酸需要量

亮氨酸是水产动物的必需氨基酸之一，亮氨酸缺乏降低草鱼、印度鲮、印度囊鳃鲇、异育银鲫等生长性能、饲料转化率及蛋白质沉积率[6,7,8,9]。近年发表的亮氨酸的需要量研究见表7-1，由此表可知鱼类对亮氨酸的需要量为1.29%～3.41%，虾蟹类对亮氨酸的需要量为1.70%～2.48%。不同品种亮氨酸需要量不同，由表7-1可知，吉富罗非鱼亮氨酸的需要量仅为1.25%，而卵形鲳鲹的需要量高达3.29%。同一品种不同规格的水产动物对亮氨酸的需要量不同，体重为2.25 g、295.85 g草鱼的亮氨酸需要量分别为1.52%、1.30%，且小规格亮氨酸需要量高于大规格需要量[6,13]。不同评价指标对亮氨酸的需要量也有较大影响，杂交石斑鱼分别以增重率、蛋白质沉积率为指标回归分析得出亮氨酸的需要量为3.25%、3.41%[10]。此外，不同的计算模型也影响亮氨酸需要量，0.38 g凡纳滨对虾分别以二次多项式模型、折线模型计算的亮氨酸需要量为2.36%、2.48%[11,12]。

表 7-1　水产动物亮氨酸需要量

类别	品种	体重（g）	模型	指标	需要量（占饲料干物质,%）	参考文献
鱼类	草鱼	295.85	二次多项式模型	增重率	1.30	Deng 等[6]
				血氨含量	1.29	
				丙二醛含量	1.32	
		2.25	折线模型	增重率	1.52	黄爱霞等[13]
				饲料转化率	1.53	
	杂交鲇	23.19	折线模型	增重率	2.81	Zhao 等[14]
	团头鲂	23.27	二次多项式模型	增重率	1.40	Liang 等[15]
				特定生长率	1.56	
		10.2	二次多项式模型	特定生长率	1.44	Ren 等[16]
				饲料转化率	1.61	
	青鱼	2.9	二次多项式模型	增重率	2.35	Wu 等[17]
				饲料转化率	2.39	
	露斯塔野鲮	0.40	二次多项式模型	增重率	1.57	Abidi 和 Khan[5]
				饲料转化率	1.55	
		0.60	二次多项式模型	增重率	1.58	Ahmed 和 Khan[7]
	杂交石斑鱼	6.92	二次多项式模型	增重率	3.25	Zhou 等[10]
				蛋白质沉积率	3.41	
	眼斑拟石首鱼	1.42	二次多项式模型	增重率	1.57	Castillo 等[18]
				蛋白质沉积率	1.63	
	吉富罗非鱼	1.94	二次多项式模型	增重率	1.25	Gan 等[19]
	卵形鲳鲹	5.76	二次多项式模型	增重率	3.28	Tan 等[20]
				特定生长率	3.29	
	印度囊鳃鲇	6.80	二次多项式模型	增重率	1.65	Farhat 等[21]
				饲料转化率	1.69	
	建鲤	7.88	二次多项式模型	增重率	1.29	伍曦[22]
	花鲈	167.82	二次多项式模型	增重率	2.76	路凯[23]
				饲料转化率	2.80	
		8.0	二次多项式模型	增重率	2.39	Li 等[24]
	吉富罗非鱼	53.65	二次多项式模型	增重率	2.33	石亚庆等[25]
				饲料转化率	2.28	
	大黄鱼	6.0	二次多项式模型	增重率	2.92	Li 等[26]
	卡特拉鲃	3.75	二次多项式模型	增重率	1.57	Zehra 和 Khan[27]
甲壳类	三疣梭子蟹	3.75	折线模型	增重率	2.21	Huo 等[28]
	凡纳滨对虾	0.38	二次多项式模型	增重率	2.36	Liu 等[11]
				饲料转化率	2.40	
		0.38	折线模型	增重率	2.48	刘福佳等[12]
		0.53	折线模型	增重率	2.46	王用黎[29]
	斑节对虾	0.02	二次多项式模型	增重率	1.70	Millamena 等[30]
	中华绒螯蟹	0.90	折线模型	特定生长率	2.36	杨霞等[31]

第三节 亮氨酸与其他氨基酸的相互作用

由于支链氨基酸在细胞膜上的转运载体相同,水产动物研究中已开展了关于亮氨酸、缬氨酸和异亮氨酸及与其他氨基酸的协同或颉颃作用的相关研究。

一、亮氨酸与缬氨酸的相互作用

Han[32]等采用2×3(亮氨酸1.6%、5.0%,缬氨酸1.2%、1.8%、2.5%)实验明确了亮氨酸能够显著影响增重率、特定生长率及饲料转化率,且与缬氨酸之间存在交互作用;低亮氨酸时,增重率和特定生长率与随着缬氨酸的增加而升高,在亮氨酸为1.6%、缬氨酸为2.5%时,增重率最高;在亮氨酸为5.0%、缬氨酸为2.5%时,增重率最低;与1.6%亮氨酸组相比,5.0%亮氨酸显著降低血细胞比容、血红蛋白、乳酸脱氢酶和谷草转氨酶活性及甘油三酯含量。亮氨酸与缬氨酸对牙鲆消化酶、免疫酶活性存在显著的交互作用,且亮氨酸为2%、缬氨酸为2.27%时,显著提高脂肪酶、超氧化物歧化酶的活性[33]。上述研究显示,饲料中高亮氨酸、高缬氨酸之间存在颉颃作用,在虹鳟、红点鲑的研究中也得到类似的结果[34,35]。

二、亮氨酸与异亮氨酸的相互作用

王莉苹[36]采用2×3(亮氨酸2.58%、5.08%,异亮氨酸1.44%、2.21%、4.44%)实验研究了亮氨酸和异亮氨酸对消化酶及非特异性免疫的影响,在亮氨酸含量为5.08%、异亮氨酸含量为1.44%时,增重率、特定生长率和蛋白酶活性显著高于低亮氨酸低异亮氨酸组;在亮氨酸含量为2.58%、异亮氨酸含量为4.44%时,脂肪酶活性显著高于低亮氨酸低异亮氨酸组,且亮氨酸和异亮氨酸之间存在显著的交互作用。此外,随着亮氨酸水平的升高,大鳞大麻哈鱼异亮氨酸的需要量也相应升高[37]。

三、亮氨酸与其他氨基酸的作用关系

眼斑拟石首鱼饲料亮氨酸含量为0.90%时血清中异亮氨酸、缬氨酸的含量显著高于亮氨酸含量为2.50%时[18]。草鱼[38]肌肉中异亮氨酸、缬氨酸的含量随着饲料中亮氨酸含量的升高而显著降低,但肌肉中总氨基酸的含量显著升高。随着饲料中亮氨酸含量的升高,杂交石斑鱼血清中缬氨酸、异亮氨酸的含量显著降低[10]。以上结果显示亮氨酸与异亮氨酸、缬氨酸存在颉颃作用,随着饲料亮氨酸含量的升高,其他两种支链氨基酸的含量显著降低。另有研究结果表明,随着饲料亮氨酸水平的升高,团头鲂血清中异亮氨酸的含量显著降低,但对缬氨酸含量的影响未达到显著水平[16]。饲料中高亮氨酸含量未对青鱼肌肉中缬氨酸、异亮氨酸的含量产生显著影响,但必需氨基酸苏氨酸、赖氨酸、亮氨酸、苯丙氨酸的含量显著升高,非必需氨基酸谷氨酸、天冬氨酸、甘氨酸、丝氨酸、胱氨酸、脯氨酸的含量同样显著升高[17]。饲料亮氨酸含量对凡纳滨对虾肌肉中异亮氨酸、缬氨酸的含量均无显著影响,但适量亮氨酸能提高苯丙氨酸、苏氨酸的含量[12]。随着饲料亮氨

酸含量的升高，中华绒螯蟹肌肉中亮氨酸和缬氨酸的含量也随之升高，丝氨酸、胱氨酸、总氨基酸、总必需氨基酸的含量也随之升高[31]。亮氨酸的吸收也受其他氨基酸的影响，肠道主要功能物质谷氨酰胺可以促进草鱼肠道对亮氨酸和脯氨酸的吸收，并显著提高肠道蛋白质的合成水平[39]。综上，饲料亮氨酸与异亮氨酸、缬氨酸、其他氨基酸的作用关系可能与品种、饲料组成等有关。

第四节　亮氨酸对水产动物蛋白质代谢的影响

亮氨酸缺乏或过量阻碍机体蛋白质的沉积。亮氨酸缺乏或过量降低凡纳滨对虾蛋白质沉积率、出肉率、肌肉蛋白质含量，且缺乏组降低肌肉总氨基酸的含量[11]。眼斑拟石首鱼饲料亮氨酸含量由0.90%提高至2.50%时，蛋白质沉积率、出肉率分别显著提高[18]。异育银鲫、团头鲂、草鱼、杂交石斑鱼，亮氨酸缺乏显著降低肝脏、肌肉 TOR 基因的表达量，抑制下游基因核糖体 S6 激酶 1（ribosome protein subunit 6 kinase 1，S6K1）的基因表达，降低蛋白质的合成代谢[9,10,16,38]。杂交石斑鱼亮氨酸缺乏或过量降低脑垂体生长激素、肝脏生长激素受体1、胰岛素样生长因子-1的基因表达量[10]。孙姝娟等[40]等注射亮氨酸24 h后，对虾 TOR 表达量是对照组的3～4倍，表明氨基酸对 TOR 的表达具有调节作用。

第五节　亮氨酸对水产动物免疫功能的影响

亮氨酸缺乏显著降低水产动物肠道、肝脏、鳃等器官的免疫功能，主要通过降低非特异性免疫相关酶活性、抗炎因子的基因表达量，降低免疫器官结构的完整性产生不良影响。亮氨酸缺乏或过量显著降低草鱼前肠、中肠、后肠溶菌酶（lysozyme activity，LA）、酸性磷酸酶（acid phosphatase activity，ACP）活性及补体C3的含量，以溶菌酶为指标，进行二次多项式回归分析得到草鱼亮氨酸需要量为1.29%；亮氨酸缺乏显著提高促炎因子白介素8（IL-8）、肿瘤坏死因子α（TNF-α）的基因表达量，降低抗炎因子白介素10（IL-10）、转化生长因子β（TGF-β）的基因表达量[41]。此外，亮氨酸缺乏或过量显著提高团头鲂肝脏 TNF-α 的基因表达量[16]。亮氨酸预处理组显著降低脂多糖（lipopolysaccharide，LPS）诱导露斯塔野鲮肝细胞的促炎因子 IL-8、TNF-α、白介素-1β（IL-1β）基因的表达量，提高抗炎因子 IL-10 的基因表达量，降低 Toll 样受体 4（toll like receptor 4，TLR4）信号通路的 TLR4、转录因子 p65（nuclear factor kappa Bp65，NF-kBp65）、髓样分化因子 88（myeloid differentiation factor 88，Myd88）、丝裂原活化蛋白激酶 p38（mitogen-activated protein kinase p38，MAPKp38）基因及蛋白的表达量[42]。上述研究表明，亮氨酸缺乏引发肠道炎症反应，影响肠道健康。亮氨酸缺乏显著降低卵形鲳鲹血清溶菌酶活性、全血血红蛋白及血细胞比容[20]。亮氨酸缺乏显著降低青鱼血清溶菌酶活性、补体C3含量；亮氨酸缺乏或过量均显著降低血液中天然抗性相关巨噬细胞蛋白（natural resistance-associated macrophage protein，NRAMP）、溶菌酶、补体C3、补体C9、干扰素-α（IFN-α）、肝杀菌肽（hepcidin，HEPC）等基因的表达量，从而影响鱼类非特异性免疫[17]。鳃不仅是鱼类呼吸、调节渗透压、酸碱平衡、氨氮排放的器官，同时也是主要的淋巴组织，具有重要的免疫功能[43,44]。亮氨酸缺乏通过细胞凋亡、紧密连接蛋白受损等破坏鳃结构的完整性，完整性受损降低免疫功能，甚至增加死亡率[45]。注射或灌喂亮氨酸显著降低罗非鱼海豚链球菌攻毒后的死亡率，主要作用途径是促进机体缬氨酸、亮氨酸、异亮氨酸的代谢及氨基酰-tRNA 的生物合成[46]。

第六节　亮氨酸对水产动物抗氧化功能的影响

活性氧自由基（reactive oxygen species，ROS）可能是引起水产动物机体氧化损伤的主要因素。机体非抗氧化酶系统和抗氧化酶系统是抑制氧化损伤的主要防御系统[47]。机体抗氧化酶系统主要受 Kelch 样环氧氯丙烷相关蛋白 α/核因子 E2 相关因子 2（Keap1α/Nrf2）信号通路调控[48]。亮氨酸缺乏或过量均显著提高肠道、肌肉丙二醛（MDA）、蛋白羰基（protein carbonyl，PC）的含量，降低谷胱甘肽（GSH）含量、铜锌超氧化物歧化酶（copper/zinc superoxide dismutase，CuZnSOD）、谷胱甘肽过氧化物酶（GPx）的活性，其作用机理可能是亮氨酸缺乏或过量通过 Keap1α/Nrf2 信号通路降低 $CuZnSOD$、GPx 的基因表达量，进而降低机体抗氧化功能[6,38]。亮氨酸缺乏或过量降低团头鲂血清 SOD、GPx、T-AOC、过氧化氢酶（CAT）的活性，提高 MDA 含量，降低 $Nrf2$、血红素氧化酶-1（heme oxygenase 1，$HO-1$）、GPx、谷胱甘肽转移酶（GST）、SOD 的基因表达量[15]。亮氨酸缺乏显著降低卵形鲳鲹血清 SOD 活性、总抗氧化能力（total antioxidant capacity，T-AOC），提高 MDA 含量[20]。亮氨酸缺乏显著降低三疣梭子蟹血清 SOD 活性，提高 MDA 含量[28]。亮氨酸含量在 1.57%～2.07%时，卵形鲳鲹血清中 T-AOC、SOD 活性显著升高，MDA 含量显著降低[49]。适量亮氨酸可以通过 Keap1α/Nrf2 信号通路提高水产动物机体抗氧化酶活性，缓解机体氧化损伤。

第七节　亮氨酸对肠道发育的影响

肠道在水产动物消化吸收、抗氧化及免疫方面发挥重要作用，且肠道消化吸收能力与肠道黏膜结构完整性具有相关性。因此，肠道发育情况直接影响水产动物生长及健康。亮氨酸缺乏显著降低罗非鱼肠道表皮生长因子（epidermal growth factor，EGF）和肠道表皮生长因子受体（epidermal growth factor receptor，EGFR）的表达量，降低肠 Na^+/K^+-ATP 酶的活性，降低胃蛋白酶、肠蛋白酶、肠脂肪酶、肠淀粉酶的活性；添加亮氨酸显著提高 EGF、EGFR 的含量，EGF 与 EGFR 结合后促进肠道的黏膜微绒毛的发育，进而提高肠道消化酶的活性和肠道的结构完整性[25]。亮氨酸缺乏显著降低青鱼 α-淀粉酶、胰蛋白酶、糜蛋白酶、弹性蛋白酶的活性[17]。此外，亮氨酸作为支链氨基酸可以为谷氨酰胺合成提供碳源和氮源，促进肠道谷氨酰胺的合成，谷氨酰胺可以为肠道提供能量，进而促进肠道的发育[39]。亮氨酸缺乏降低肠消化酶活性的主要原因可能是由于肠道发育受到抑制。亮氨酸缺乏显著降低卵形鲳鲹肠道微绒毛数量、长度，提高隐窝深度[20]。亮氨酸缺乏或过量显著降低杂交石斑鱼褶皱高度、褶皱宽度、肠上皮细胞高度、微绒毛高度[10]。

亮氨酸含量在1.57%～2.07%时，显著提高卵形鲳鲹肠淀粉酶、胃蛋白酶的活性[49]。亮氨酸促进肠道完整性主要是通过提高肠道紧密连接蛋白 claudin b、claudin c、claudin 3、claudin 15、occludin、ZO-1 等基因的表达量，进而促进肠道发育、提高消化酶的活性[41]。

第八节　小结与展望

近年来已经开展了大量亮氨酸的相关研究，确定了鱼类对亮氨酸的需要量为1.29%～3.41%，虾蟹类对亮氨酸的需要量为1.70%～2.48%；开展了亮氨酸与缬氨酸、异亮氨酸氨基酸相互作用的研究；探讨了亮氨酸对蛋白质代谢、免疫功能、抗氧化功能的影响。未来对于亮氨酸的研究，应不局限于鱼的小规格阶段，需拓展至生长中后期；作为支链氨基酸，其与其他氨基酸和营养物质的互作关系亦是值得深入研究的方向，从而更全面的了解亮氨酸的功能以及营养作用机制，为水产养殖业、动物健康保障和功能性物质的开发奠定基础。

参考文献

[1] Li P, Mai K S, Trushenski J, et al. New developments in fish amino acid nutrition: towards functional and environmentally oriented aquafeeds [J]. Amino Acids, 2009, 37: 43-53.

[2] Li P, Yin Y L, Li D, et al. Amino acids and immune function [J]. British Journal of Nutrition, 2007, 98: 237-252.

[3] Lynch C J, Adams S H. Branched-chain amino acids in metabolic signalling and insulin resistance [J]. Nature Reviews Endocrinology, 2014, 10: 723-736.

[4] Meijer A J, Dubbelhuis P F. Amino acid signalling and the integration of metabolism [J]. Biochemical and Biophysical Research Communications, 2004, 313: 397-403.

[5] Abidi S F, Khan M A. Dietary leucine requirement of fingerling Indian major carp, *Labeo rohita* (Hamilton) [J]. Aquaculture Research, 2007, 38: 478-486.

[6] Deng Y P, Jiang W D, Liu Y, et al. Differential growth performance, intestinal antioxidant status and relative expression of Nrf2 and its target genes in young grass carp (*Ctenopharyngodon idella*) fed with graded levels of leucine [J]. Aquaculture, 2014, 434: 66-73.

[7] Ahmed I, Khan M A. Dietary branched-chain amino acid valine, isoleucine and leucine requirements of fingerling Indian major carp, *Cirrhinus mrigala* (Hamilton) [J]. British Journal of Nutrition, 2006, 96: 450-460.

[8] Khan M A. Response of fingerling stinging catfish, *Heteropneustes fossilis* (Bloch) to varying levels of dietary L-leucine in relation to growth, feed conversion, protein utilization, leucine retention and blood parameters [J]. Aquaculture Nutrition, 2014, 20: 291-302.

[9] Zou T, Cao S P, Xu W J, et al. Effects of dietary leucine levels on growth, tissue protein content and relative expression of genes related to protein synthesis in juvenile gibel carp (*Carassius auratus gibelio* var. CAS Ⅲ) [J]. Aquaculture Research, 2018, 49: 2240-2248.

[10] Zhou Z Y, Wang X, Wu X Y, et al. Effects of dietary leucine levels on growth, feed utilization,

neuro-endocrine growth axis and TOR-related signaling genes expression of juvenile hybrid grouper (*Epinephelus fuscoguttatus* ♀×*Epinephelu slanceolatus* ♂) [J]. Aquaculture, 2019, 04: 172-181.

[11] Liu F J, Liu Y J, Tian L X, et al. Quantitative dietary leucine requirement of juvenile Pacific white shrimp, *Litopenaeus vannamei* (Boone) reared in low-salinity water [J]. Aquaculture Nutrition, 2014, 20 (3): 332-340.

[12] 刘福佳, 李雪菲, 刘永坚, 等. 低盐度条件下的凡纳滨对虾幼虾亮氨酸营养需求 [J]. 中国水产科学, 2014, 21 (5): 963-972.

[13] 黄爱霞, 孙丽慧, 陈建明, 等. 饲料亮氨酸水平对幼草鱼生长、饲料利用及体成分的影响 [J]. 饲料工业, 2018, 39 (2): 26-32.

[14] Zhao Y, Li J Y, Jiang Q, et al. Leucine Improved Growth Performance, Muscle Growth, and Muscle Protein Deposition Through AKT/TOR and AKT/FOXO3a Signaling Pathways in Hybrid Catfish *Pelteobagrus vachelli* × *Leiocassis longirostris* [J]. Cells, 2020, 9 (2): 1-22.

[15] Liang H, Mokrani A, Ji K, et al. Dietary leucine modulates growth performance, Nrf2 antioxidant signaling pathway and immune response of juvenile blunt snout bream (*Megalobrama amblycephala*) [J]. Fish & Shellfish Immunology, 2018, 73: 57-65.

[16] Ren M C, Habte-michael H T, Liu B, et al. Dietary leucine level affects growth performance, whole body composition, plasma parameters and relative expression of TOR and TNF-a in juvenile blunt snout bream, *Megalobrama amblycephala* [J]. Aquaculture, 2015, 448: 162-168.

[17] Wu C L, Chen L, Lu Z B, et al. The effects of dietary leucine on the growth performances, body composition, metabolic abilities and innate immune responses in black carp *Mylopharyngodon piceus* [J]. Fish & Shellfish Immunology, 2017, 67: 419-428.

[18] Castillo S, Gatlin D M. Dietary requirements for leucine, isoleucine and valine (branched-chain amino acids) by juvenile red drum *Sciaenops ocellatus* [J]. Aquaculture Nutrition, 2018, 24: 1056-1065.

[19] Gan L, Zhou L L, Li X X, et al. Dietary leucine requirement of Juvenile Nile tilapia, *Oreochromis niloticus* [J]. Aquaculture Nutrition, 2016, 22 (5): 1040-1046.

[20] Tan X, Lin H, Huang Z, et al. Effects of dietary leucine on growth performance, feed utilization, non-specific immune responses and gut morphology of juvenile golden pompano *Trachinotus ovatus* [J]. Aquaculture, 2016, 465 (1): 100-107.

[21] Farhat Khan M A. Response of fingerling stinging catfish, *Heteropneustes fossilis* (Bloch) to varying levels of dietary l-leucine in relation to growth, feed conversion, protein utilization, leucine retention and blood parameters [J]. Aquaculture Nutrition, 2014, 20 (3): 291-302.

[22] 伍曦. 亮氨酸对幼建鲤生长性能和免疫功能的影响 [D]. 雅安: 四川农业大学, 2011

[23] 路凯. 花鲈对亮氨酸、异亮氨酸和缬氨酸需要量的研究 [D]. 青岛: 中国海洋大学, 2015.

[24] Li Y, Cheng Z Y, Mai K S, et al. Dietary Leucine Requirement of Juvenile Japanese Seabass (*Lateolabrax Japonicus*) [J]. Journal of Ocean University of China, 2015, 14 (1): 121-126.

[25] 石亚庆, 孙玉轩, 罗莉, 等. 吉富罗非鱼亮氨酸需要量研究 [J]. 水产学报, 2014, 38 (10): 1778-1785.

[26] Li Y, Ai Q H, Mai K S, et al. Dietary leucine requirement for juvenile large yellow croaker *Pseudosciaena crocea* (Richardson, 1846) [J]. Journal of Ocean University of China, 2010, 9 (4): 371-375.

[27] Zehra S, Khan M A. Dietary leucine requirement of fingerling *Catla catla* (Hamilton) based on

[27] (continued) growth, feed conversion ratio, RNA/DNA ratio, leucine gain, blood indices and carcass composition [J]. Aquaculture International, 2015, 23: 577-595.

[28] Huo Y W, Jin M, Sun P, et al. Effect of dietary leucine on growth performance, hemolymph and hepatopancreas enzyme activities of swimming crab, *Portunus trituberculatus* [J]. Aquaculture Nutrition, 2017, 23: 1341-1350.

[29] 王用黎. 凡纳滨对虾幼虾对苏氨酸、亮氨酸、色氨酸和缬氨酸需要量的研究 [D]. 湛江: 广东海洋大学, 2013.

[30] Millamena O M, Teruel M B, Kanazawa A, et al. Quantitative dietary requirements of postlarval tiger shrimp, *Penaeus monodon*, for histidine, isoleucine, leucine, phenylalanine and tryptophan [J]. Aquaculture, 1999, 179: 169-179.

[31] 杨霞, 叶金云, 周志金, 等. 中华绒螯蟹幼蟹对亮氨酸和异亮氨酸的需要量 [J]. 水生生物学报, 2014, 38 (06): 1062-1070.

[32] Han Y Z, Han R Z, Koshio S, et al. Interactive effects of dietary valine and leucine on two sizes of Japanese flounder *Paralichthys olivaceus* [J]. Aquaculture, 2014, 432: 130-138.

[33] 王旭, 周婧, 薛晓强, 等. 牙鲆饲料中异亮氨酸与缬氨酸的交互作用对消化酶和部分免疫酶的影响 [J]. 饲料工业, 2018, 39 (20): 23-28.

[34] Yamamoto T, Shima T, Furuita H. Antagonistic effects of branched-chain amino acids induced by excess protein-bound leucine in diets for rainbow trout (*Oncorhynchus mykiss*) [J]. Aquaculture, 2004, 232: 539-550.

[35] Hughes S G, Rumsey G L, Nesheim M C. Effects of dietary excesses of branched chain amino acids on the metabolism and tissue composition of lake trout (*Salvelinus namaycush*) [J]. Comparative Biochemistry and Physiology A-Molecular & Integrative Physiology, 1984, 78: 413-418.

[36] 王莉苹. 牙鲆 (*Paralichthys olivaceus*) 饲料中亮氨酸与异亮氨酸交互作用的研究 [D]. 大连: 大连海洋大学, 2017.

[37] Chance R E, Mertz E T. Halver J E. Nutrition of salmonoid fishes. XII. Isoleucine, leucine, valine and phenylalanine requirements of chinook salmon and interrelations between isoleucine and leucine for growth [J]. Journal of Nutrition, 1964, 83: 177-185.

[38] Deng Y P, Jiang W D, Liu Y, et al. Dietary leucine improves flesh quality and alters mRNA expressions of Nrf2-mediated antioxidant enzymes in the muscle of grass carp (*Ctenopharyngodon idella*) [J]. Aquaculture, 2016, 452: 380-387.

[39] 叶元土, 王永玲, 蔡春芳, 等. 谷氨酰胺对草鱼肠道L-亮氨酸、L-脯氨酸吸收及肠道蛋白质合成的影响 [J]. 动物营养学报, 2007, 19 (1): 28-32.

[40] 孙姝娟, 刘梅, 彭劲松, 等. 中国明对虾TOR基因的克隆及精氨酸、亮氨酸对其表达的影响 [J]. 海洋科学, 2010, 34 (6): 71-80.

[41] Jiang W D, Deng Y P, Liu Y, et al. Dietary leucine regulates the intestinal immune status, immune-related signalling molecules and tight junction transcript abundance in grass carp (*Ctenopharyngodon idella*) [J]. Aquaculture, 2015, 444: 134-142.

[42] Giri S S, Sen S S, Jun J W, et al. Protective effects of leucine against lipopolysaccharide-induced inflammatory response in *Labeo rohita* fingerlings [J]. Fish & Shellfish Immunology, 2016, 52: 239-247.

[43] Evans D H, Piermarini P M, Choe K P. The multifunctional fish gill: dominant site of gas exchange, osmoregulation, acid-base regulation, and excretion of nitrogenous waste [J]. Physiological Reviews,

2005, 85 (1): 97-177.

[44] Martins C, De Matos A P A, Costa M H, et al. Alterations in juvenile flatfish gill epithelia induced by sediment-bound toxicants: a comparative *in situ* and *ex situ* study [J]. Environmental Research, 2015, 112: 122-130.

[45] Jiang W D, Deng Y P, Zhou X Q, et al. Towards the modulation of oxidative damage, apoptosis and tight junction protein in response to dietary leucine deficiency: A likely cause of ROS-induced gill structural integrity impairment [J]. Fish & Shellfish Immunology, 2017, 70: 609-620

[46] Ma Y M, Yang M J, Wang S Y, et al. Liver functional metabolomics discloses an action of L-leucine against *Streptococcus iniae* infection in tilapias [J]. Fish & Shellfish Immunology, 2015, 45: 414-421.

[47] Martinez-Alvarez R M, Morales A E, Sanz A. Antioxidant defenses in fish: biotic and abiotic factors [J]. Reviews in Fish Biology and Fisheries, 2005, 15: 75-88.

[48] Giulianl M E, Regoli F. Identification of the Nrf2-Keap1 pathway in the European eel *Anguilla anguilla*: role for a transcriptional regulation of antioxidant genes in aquatic organisms [J]. Aquatic Toxicology, 2014, 150: 117-123.

[49] 黄忠, 周传朋, 林黑着, 等. 饲料异亮氨酸水平对卵形鲳鲹消化酶活性和免疫指标的影响 [J]. 南方水产科学, 2017, 13 (1): 50-57.

第八章 异亮氨酸

异亮氨酸是支链氨基酸中唯一的生糖兼生酮氨基酸[1,2]，与缬氨酸和亮氨酸一样具有由甲基侧链形成的分支结构[3]，由 Ehrlich 发现并提取，其理化性质与亮氨酸差别较大，但化学组成相同，结构式为"α-氨基-β-甲基戊酸"被认为是有别于亮氨酸的一种氨基酸，故将其定名为异亮氨酸。异亮氨酸由于在 α 位和 β 位具有两个不对称碳原子，因而存在 D、L、D 别和 L 别四种旋光异构体，但 L-异亮氨酸是其 4 种立体异构体中唯一一个在自然界存在的，其余三种均无营养价值。在机体的生命活动中，异亮氨酸起重要的作用，能够促进蛋白质的合成，并抑制蛋白质的分解；促进机体合成需要的激素或酶类，如胰岛素和生长激素等[4]；此外，作为谷氨酰胺合成的底物，调节氨基酸代谢，为机体提供能量[5]。本章综述了异亮氨酸在水产动物营养与代谢中的研究进展，包括水产动物异亮氨酸的需要量及其在蛋白质代谢、免疫、抗氧化功能、肠道发育和肉质等方面的作用，为异亮氨酸在水产动物饲料中的应用提供借鉴。

第一节 异亮氨酸的理化性质

异亮氨酸（I-isoleucine，Ile），分子式为 $C_6H_{13}NO_2$，分子质量 131.17，分子结构如图 8-1 所示，熔点为 284 ℃，沸点为 168~170 ℃。外观为白色结晶小片或结晶性粉末，在乙醇中结晶可形成菱形叶片或片状晶体。无臭，略有苦味。溶于水，微溶于乙醇。

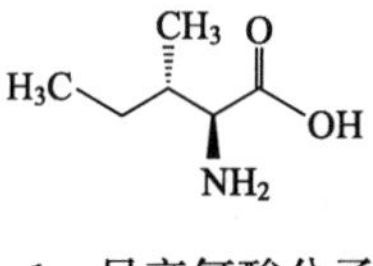

图 8-1 异亮氨酸分子结构

第二节 水产动物异亮氨酸需要量

异亮氨酸作为必需氨基酸之一，饲料中异亮氨酸的缺乏会降低水产动物的生长性能和蛋白质效率，增加饲料系数[6,7]。基于生长试验研究的水产动物对异亮氨酸的需要量见表 8-1，由表可知，因品种、规格及评价指标的不同而不同，鱼类对异亮氨酸的需要量为 0.62%~1.98%，虾蟹类对异亮氨酸的需要量为 1.01%~2.25%，而仿刺参对异亮氨酸的需要量为 0.98%。不同品种的水产动物对异亮氨酸的需要量差别较大，斑点叉尾鮰对异亮氨酸的需要量仅为 0.62%[8]，而中华绒螯蟹对异亮氨酸的需要量则为 2.25%[9]。而同一品种不同规格的水产动物对异亮氨酸的需要量也不同，体重为 8.25 g 和 256.80 g 草鱼对异亮氨酸的需要量分别为 1.46% 和 1.21%[10,11]；体重为 0.61 g 和 1.2 g 的卡特拉鲃对异亮氨酸的需要量分别为 1.13% 和 0.94%[12,13]；而小规格的尼罗罗非鱼（0.087 g）对异亮氨酸的需要量低于大规格（1.24 g）的尼罗罗非鱼[14,15]。评价指标的不同对异亮氨酸的需要量也有一定的影响，0.61 g 的印度鲮[6]经二次多项式模型分析后，评价指标为增重率、蛋白质效率和饲料系数时，得出异亮氨酸的需要量分别为 1.32%、1.21% 和 1.23%；大黄鱼[16]经二次多项式模型分析后，以增重率和饲料系数为评价指标，得出异亮氨酸的需要量分别为 1.71% 和 1.59%。

表 8-1 水产动物异亮氨酸需要量

品种	体重（g）	模型	指标	需要量（%）		参考文献
				占饲料干物质	占饲料蛋白	
鱼类						
草鱼	8.25	二次多项式模型	增重率	1.46	4.14	尚晓迪等[10]
			特定生长率	1.49	4.23	
			饲料系数	1.45	4.11	
			蛋白质效率	1.41	4.0	
	256.80	二次多项式模型	增重率	1.21	3.93	Gan 等[11]
	256.80	二次多项式模型	蛋白羧基	1.33	4.30	Feng 等[17]
			claudin-3 mRNA 水平	1.10	3.54	
团头鲂	10.20	二次多项式模型	特定增长率	1.38	4.06	Ren 等[18]
			饲料系数	1.40	4.12	
建鲤	6.90	二次多项式模型	特定生长率	1.29	3.84	Zhao 等[19]
	6.90	二次多项式模型	溶菌酶活性，丙二醛	1.25~1.33	3.71~3.96	Zhao 等[20]
	6.90	二次多项式模型	肠道健康	0.98~1.01	2.90~3.02	Zhao 等[21]
印度鲮	0.61	二次多项式模型	增重率	1.32	3.15	Ahmed 等[6]
			饲料系数	1.23		
			蛋白质效率	1.21		
	0.93~0.95	折线模型	增重率	1.25	3.12	Benakappa 等[22]
露斯塔野鲮	0.40	二次多项式模型	增重率	1.59	3.98	Khan 等[7]
			饲料系数	1.53	3.93	
			蛋白质效率	1.52	3.80	
			特定增长率	1.58	3.95	
			蛋白质沉积率	1.57	3.93	
卡特拉鲃	0.61	二次多项式模型	增重率	1.13	3.42~3.58	Zehra 等[12]
			蛋白质效率	1.18		
			Ile 沉积率	1.13		
			RNA/DNA	1.16		
			体蛋白	1.14		
	1.20	折线模型	增重率	0.94	2.35	Ravi 等[13]

(续)

品种	体重（g）	模型	指标	需要量（%） 占饲料干物质	需要量（%） 占饲料蛋白	参考文献
鲈	8.00	二次多项式模型	特定生长率	1.94	4.69	李燕[16]
			饲料效率	1.97	4.69	
	159.33	二次多项式模型	增重率	1.88	4.41	路凯等[23]
			蛋白质沉积率	1.84	4.32	
大黄鱼	6.00	二次多项式模型	特定生长率	1.71	3.98	李燕[16]
			饲料效率	1.59	3.70	
尼罗罗非鱼	0.087	折线模型	增重率	0.87	3.11	Santiago等[14]
	2.46	折线模型	增重率	1.37	5.00	FA等[15]
	18.09	多重比较	增重率	0.70	2.50	Neu等[24]
吉富罗非鱼	49.12	二次多项式模型	增重率	1.47	4.52	孙玉轩[25]
			特定生长率	1.47	4.52	
			蛋白质效率	1.44	4.43	
			蛋白质沉积率	1.44	4.43	
			饲料系数	1.43	4.40	
遮目鱼	0.32	折线模型	增重率	1.60	4.00	Borlongan等[26]
乌鳢	6.74	二次多项式模型	增重率	1.81	4.01	Sharf等[27]
			饲料系数	1.82	4.04	
			蛋白质沉积率	1.84	4.09	
			异亮氨酸沉积率	1.80	3.99	
美国红鱼	3.68	二次多项式模型	增重率	1.11	3.00	Castillo等[28]
红鳍东方鲀	29.00	二次多项式模型	蛋白质沉积率	1.93		孙志远等[29]
			鱼体粗蛋白含量	1.83		
			肌肉Ile含量	1.96		
大鳞大麻哈鱼	0.98	折线模型	增重率	0.90～1.10	2.20～2.70	Chance等[30]
杂交石斑鱼	6.00	二次多项式模型	增重率	1.98	3.99	Zhou等[31]
虹鳟	47.00	二次多项式模型	增重率	1.40	4.00	Rodehutscord等[32]
湖鳟	3.20	多重比较	增重率	0.72	2.06	Steven等[33]
斑点叉尾鮰	200.00	折线模型	增重率	0.62	2.58	Wilson等[8]
金昌鱼	6.36	二次多项式模型	增重率	1.74	4.04～4.07	Huang等[34]
			饲料系数	1.74		
			蛋白质效率	1.75		
虾蟹类						
中华绒螯蟹	0.90	折线模型	特定生长率	2.25	5.72	杨霞等[9]

(续)

品种	体重（g）	模型	指标	需要量（%）		参考文献
				占饲料干物质	占饲料蛋白	
斑节对虾	0.02	折线模型	增重率	1.01	2.70	Millamena 等[35]
南美白对虾	0.43	二次多项式模型	增重率	1.59	3.88	Liu 等[36]
其他						
仿刺参	9.20	折线模型	增重率	0.98	5.06	韩秀杰[37]

根据 Kaushik 等[38]的方法，现将以上水产动物的异亮氨酸需要量进行了 Meta 分析。如图 8-2 所示，将增重率进行标准化响应处理后与异亮氨酸需要量进行回归分析，异亮氨酸需要量（占饲料干物质百分比）的二次多项式回归分析模型为，$y_t = -43.131x^2 + 130.21x + 1.5149$，$R^2 = 0.9746$；异亮氨酸需要量（占饲料蛋白百分比）的二次多项式回归分析模型为，$y = -5.3195x^2 + 45.502x + 0.5164$，$R^2 = 0.9815$。以增重率为评价指标，经计算得到水产动物对异亮氨酸的需要量为占饲料干物质的 1.51%，占饲料蛋白质的 4.28%。

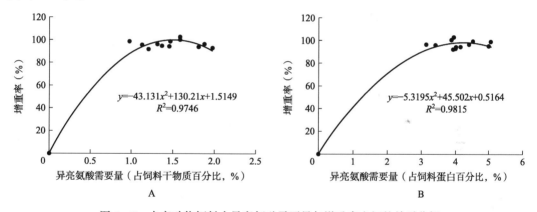

图 8-2 水产动物饲料中异亮氨酸需要量与增重率之间的关系分析

此外，关于氨基酸需要量的研究方法，除了比较常用的利用生长试验测定的氨基酸需要量以外，还有氨基酸氧化法、析因法、氮平衡法等[39]。在水产动物上，近年来 Hua[40]等利用析因法研究了大西洋鲑和虹鳟对异亮氨酸的需要量。Walton 等[41]利用氨基酸氧化法计算了赖氨酸的需要量，而 Kim 等[42]认为该方法不适合对虹鳟苯丙氨酸需要量的测定，而利用该方法检测水产动物异亮氨酸需要量的文献还未见报道。研究者利用氮平衡方法研究了吉富罗非鱼的异亮氨酸需要量为占饲料蛋白的 3.26%[43]。此外，利用全鱼氨基酸组成或肌肉氨基酸组成来评估必需氨基酸需要量的方法，具有节省时间和成本的优势，通常以 A/E 值作为评价指标（A/E=单一必需氨基酸量/必需氨基酸总量×1 000）。通过该方法测定的不同品种的鱼对异亮氨酸的需要量为 0.47%～1.4%（占饲料干物质），2.00%～3.44%（占饲料蛋白）（表 8-2）。

表 8-2 水产动物异亮氨酸需要量（其他方法）

品种	体重（g）	方法	指标	需要量 占饲料干物质百分比（%）	需要量 占饲料蛋白百分比（%）	参考文献
非洲龙鱼	0.20	卵/全鱼氨基酸组成	A/E值	—	2.90	Monentcham 等[44]
非洲龙鱼	2.00~60.00	卵/全鱼氨基酸组成	A/E值	—	2.90	Monentcham 等[44]
非洲龙鱼	400.00	卵/全鱼氨基酸组成	A/E值	—	2.90	Monentcham 等[44]
非洲龙鱼	4 000.00~5 000.00	卵/全鱼氨基酸组成	A/E值	—	2.90	Monentcham 等[44]
北极红点鲑	20.0~35.0	全鱼氨基酸组成	A/E值	1.20	—	Gurure 等[45]
大西洋鲑	0.2~20.0	析因法	—	1.24~1.56	3.4~3.6	Hua 等[40]
大西洋鲑	20.0~100	析因法	—	1.23~1.55	3.4~3.5	Hua 等[40]
大西洋鲑	100~500	析因法	—	1.20~1.51	3.3~3.4	Hua 等[40]
大西洋鲑	500~1 500	析因法	—	1.11~1.39	3.1~3.2	Hua 等[40]
大西洋鲑	1 500~2 500	析因法	—	0.96~1.21	2.7~2.8	Hua 等[40]
条纹鲮脂鲤	—	全鱼氨基酸组成	A/E值	—	2.92	De Almeida Bicudo 等[46]
条纹丽脂鲤	16.20	全鱼氨基酸组成	A/E值	1.40	3.28	Furuya 等[47]
鲤	76.40	全鱼氨基酸组成	A/E值	0.80	2.30	Ogino[48]
欧洲鲈	—	全鱼氨基酸组成	A/E值	—	2.60	Kaushik[49]
金头鲷	—	全鱼氨基酸组成	A/E值	—	2.60	Kaushik[49]
金头鲷	2.20	全鱼氨基酸组成	A/E值	0.47	1.12	Gaber 等[50]
金头鲷	4.60	全鱼氨基酸组成	A/E值	1.00	2.55	Peres 等[51]
银鲳	25.03	全鱼/肌肉氨基酸组成	A/E值	—	1.61~1.78	Tesser 等[52]
牙鲆	3.00	全鱼氨基酸组成	A/E值	1.00	2.00	Forster[53]
亚马逊石脂鲤	—	全鱼氨基酸组成	A/E值	—	3.08	De Almeida Bicudo 等[46]
尼罗罗非鱼	4.31	全鱼氨基酸组成	A/E值	1.15	3.44	Ogunji 等[54]
尼罗罗非鱼	20.40	全鱼氨基酸组成	A/E值	0.88	2.84	Diógenes 等[55]
吉富罗非鱼	158.78	氮平衡	蛋白质效率	—	3.26	Do Nascimento 等[43]
锯腹脂鲤	54.30	全鱼氨基酸组成	A/E值	0.82	—	Khan 等[56]
锯腹脂鲤	822.50	全鱼氨基酸组成	A/E值	0.81	—	Khan 等[56]
锯腹脂鲤	1 562.80	全鱼氨基酸组成	A/E值	0.85	—	Khan 等[56]
贝尔鲑脂鲤	—	全鱼氨基酸组成	A/E值	—	2.40~2.45	Kasozi 等[57]
虹鳟	0.2~20.0	析因法	—	1.09~1.37	3.0~3.1	Hua 等[40]
虹鳟	20.0~100	析因法	—	1.07~1.35	3.0~3.1	Hua 等[40]
虹鳟	100~500	析因法	—	0.99~1.25	2.8	Hua 等[40]
虹鳟	500~1 500	析因法	—	0.75~0.95	2.1~2.2	Hua 等[40]
虹鳟	126	全鱼氨基酸组成	A/E值	0.9	2.4	Ogino[48]

(续)

品种	体重（g）	方法	指标	需要量 占饲料干物质百分比（%）	需要量 占饲料蛋白百分比（%）	参考文献
美国红鱼	—	肌肉氨基酸组成	A/E 值	1.00	2.90	Moon 等[58]
真鲷	1.70	全鱼氨基酸组成	A/E 值	1.10	2.20	Forster 等[53]
银鲳	—	全鱼氨基酸组成	A/E 值	—	2.00	Hossain 等[59]
条纹鲈	111.00 / 790.00	全鱼氨基酸组成	A/E 值	0.90	2.50	Small 等[60]
大菱鲆	—	全鱼氨基酸组成	A/E 值	—	2.60	Kaushik[49]
高首鲟	66.70	全鱼氨基酸组成	A/E 值	1.30	2.99	Ng 等[61]

第三节 异亮氨酸与其他支链氨基酸的相互作用

BCAA 在通过细胞膜时会竞争载体，因而饲料中 BCAA 的失衡会导致动物生长性能或免疫功能的下降[62]。关于 BCAAs 的颉颃作用，在单胃和瘤胃陆生动物中已经进行了报道[63]，但其在水生动物中相互作用的结果却没有统一结论[64]。对斑点叉尾鲴[8]的研究表明，血清中缬氨酸和异亮氨酸含量在亮氨酸缺乏的试验组中显著增加。另外，饲料中异亮氨酸或缬氨酸的增加引起了血清中 BCAAs 水平的变化。对大鳞大麻哈鱼的研究表明，其对异亮氨酸的需要随着饲料中亮氨酸含量的增加而增加[30]。在对牙鲆的研究中发现，在高异亮氨酸试验组中，牙鲆的增重率随缬氨酸水平的升高而显著升高，在低异亮氨酸组中没有该现象[65]；而在亮氨酸与异亮氨酸的交互作用中研究发现，以生长为评价指标，在低水平（2.58%）的亮氨酸饲料中，牙鲆幼鱼的生长性能随着异亮氨酸的添加水平提高而显著提高，但在高水平（4.44%）亮氨酸饲料中添加异亮氨酸其生长性能随着添加量的增加而显著下降，产生了颉颃作用；生长性能表现最佳的组为低亮氨酸-高异亮氨酸组和高亮氨酸-低亮氨酸组[66]。在虹鳟[41]和湖鳟[33]的研究中发现，异亮氨酸和亮氨酸存在颉颃作用。但 Choo 等[67]认为过量亮氨酸对虹鳟生长的抑制和形态异常不是由于 BCAA 的颉颃作用，而是由于过量亮氨酸本身的毒性作用。在斑点叉尾鲴中，当饲料中 3 种 BCAA 满足要求时，BCAA 之间没有发生颉颃作用[68]，在尼罗罗非鱼的研究中也发现同样的现象[15]。目前研究表明，尼罗罗非鱼的 BCAA 最优比值是异亮氨酸：亮氨酸：缬氨酸＝1：1.3：0.9[15]，印度鲮[6]的 BCAA 最优比值是异亮氨酸：亮氨酸：缬氨酸＝1：1.2：1.2，杂交石斑鱼的 BCAA 最优比值是异亮氨酸：亮氨酸：缬氨酸＝1：1.7：1.4[31]。

第四节 异亮氨酸的生理功能

一、异亮氨酸对蛋白质代谢的影响

氨基酸在调节蛋白质合成中起着至关重要的作用，而一种必需氨基酸的缺失可能会损害多种生理功能[38,69]。研究表明，异亮氨酸作为调节蛋白质转换的信号分子，在蛋白质代谢中起着重要的作用[70]，这个结果在尼罗罗非鱼中也得到验证[15]，添加适宜水平的异亮氨酸显著提高其蛋白质效率和氨基酸的沉积率，同时上调成肌因子 *MyoG* 和 *MyoD* 基因的表达量，MyoG 和 MyoD 作为转录因子在控制鱼类肌肉生长起关键作用[71]。此外，研究证实异亮氨酸通过调节鱼体内 mTOR 活性来调控蛋白质的合成[72]。喂食富含异亮氨酸的饲料显著增加五条鰤[73]、印度鲮[6]、露斯塔野鲮[7]和美国红鱼[28]蛋白质的合成。饲料中添加 1.23% 和 1.47% 异亮氨酸可以分别提高红鳍东方鲀[29]肌肉和血清中多种氨基酸含量，而在 1.93% 异亮氨酸水平下显著提高了蛋白质的沉积率。同样，饲料中添加适宜水平异亮氨酸可以显著提高中华绒螯蟹肌肉中多种氨基酸的含量[9]。异亮氨酸缺乏会显著降低草鱼的生长性能，而肌肉中的蛋白质和脂肪的沉积随着饲料中异亮氨酸的增加而增加[74]。黄康等[75]研究了异亮氨酸对鳜的蛋白质代谢的影响，发现其能够激活 mTOR 信号通路，提高谷氨酸脱氢酶基因、谷草转氨酶基因和腺苷酸脱氨酶等氮代谢基因的表达水平，从而促进氨基酸代谢和氨氮排泄增多，氨氮的变化可反映鱼体内蛋白质的转化与利用情况[76]，因此证实了异亮氨酸可以促进蛋白质的沉积与利用。

二、异亮氨酸对机体免疫功能的影响

异亮氨酸在提高非特异性免疫细胞功能和免疫因子活性方面起到了重要的作用。非特异性免疫细胞包括树突状细胞、吞噬细胞和杀伤细胞等，免疫因子包括补体（C）、酶类物质、白细胞介素（IL）等，它们在抵挡入侵的病原体中发挥着重要的作用[77]。异亮氨酸可以增强吞噬细胞的吞噬和杀菌能力，在建鲤的研究中表明，其能提高头肾指数，提高血清中白细胞数量、白细胞的吞噬能力和 IgM 的含量[20]；在牙鲆的研究中表明，适宜水平的异亮氨酸可以提高头肾的呼吸爆发活性[78]。研究表明，适宜水平的异亮氨酸能够提高牙鲆[78]、吉富罗非鱼[25]、卵形鲳鲹[79]、建鲤[20]的溶菌酶、酸性磷酸酶活性，提高 C3 和 C4 含量，从而提高机体非特异性免疫功能。目前关于异亮氨酸对鱼类炎症因子的影响仅有少量报道。在草鱼的研究中发现，异亮氨酸显著抑制了其鳃中由巨噬细胞分泌的促炎因子——肿瘤坏死因子-α（*TNF-α*）、白细胞介素-1β（*IL-1β*）、白细胞介素-8（*IL-8*）基因的表达，显著促进了抗炎因子白细胞介素-10（*IL-10*）和转化生长因子-β（*TGF-β*）基因的表达[17]。Zhao 等[20,21]对建鲤的研究表明，饲料中添加适宜水平的异亮氨酸可以改善建鲤的免疫反应，促进肠道和头肾 TGF-β 的表达，抑制 TNF-α 的表达，同时发现，异亮氨酸可以提高草鱼 mTOR 和其下游效应蛋白核糖体 S6 激酶（ribosomal protein S6 kinase，*S6K1*）mRNA 的表达，研究表明 mTOR 活性的上调会抑制 TNF-α 含量的增加[17,80]，而 S6K1 则能激活 IL-10 和 TGF-β[81]。由此，异亮氨酸很可能是通过上调 *mTOR* 和 *S6K1* 基因表达来调控促炎因子和抗炎因子的表达，从而调节机体的抗炎作用，但具体的调控机制还有待深入研究。

三、异亮氨酸对机体抗氧化功能的影响

机体在病理情况下产生过多的活性氧（ROS）会导致氧化损伤[77]，而衡量氧化损伤程度高低的指标包括一些氧化产物和上皮细胞结构的完整性[82]。在建鲤[21]、草鱼[17]、卵

形鲳鲹[79]和建鲤肠道上皮细胞[82]中均发现适宜水平的异亮氨酸能显著下调肠道组织或肠细胞以及血清中的蛋白质的氧化产物——蛋白质羰基（PC）和脂肪的氧化产物——丙二醛（MDA）含量，以及下调评价上皮细胞完整性的胞外乳酸脱氢酶（lactate dehydrogenase，LDH）的活性，从而抵御机体的氧化应激。谷胱甘肽（GSH）能有效地清除ROS，谷氨酸半胱氨酸连接酶（glutamate cysteine ligase，GCL）是GSH生物合成过程中的限速酶。在建鲤[21]和草鱼[17]的研究中发现，饲料中补充异亮氨酸可以显著提高GSH含量，同时提高草鱼肌肉 GCL 的基因表达水平。

一些抗氧化物酶可以清除机体内的ROS。研究表明，添加适宜水平的异亮氨酸可以显著增加牙鲆超氧化物歧化酶（SOD）活性[78]，抑制过氧化氢酶（CAT）活性[83]；提高吉富罗非鱼[25]和卵形鲳鲹[79]血清中SOD活性；提高建鲤[21]肠道谷胱甘肽过氧化物酶（GPx）、SOD、谷胱甘肽硫转移酶（GST）、CAT和谷胱甘肽还原酶（GR）酶活性，提高草鱼鳃SOD、CAT和GPx酶活性。

此外，在草鱼的研究中发现，异亮氨酸缺乏或过量会导致草鱼肌肉中核因子E2相关因子2（nuclear factor E2-related factor 2，Nrf2）mRNA表达水平的下降，适宜的异亮氨酸可抑制其降低[17]，Nrf2作为重要的转录因子可调节抗氧化酶基因的表达[84]，这说明异亮氨酸是通过Nrf2信号通路激活抗氧化防御。

四、异亮氨酸对肠道健康的影响

肠道消化酶活性是评价肠道对营养物质吸收能力的重要指标，除前面所述的异亮氨酸对肠道抗氧化和免疫指标的影响外[21]，适宜的异亮氨酸水平可以显著提高吉富罗非鱼[25]、卵形鲳鲹[79]肠道的消化酶活性。王莉苹等[83]和王旭等[85]对牙鲆的研究发现异亮氨酸和亮氨酸或异亮氨酸和缬氨酸的交互作用均可显著提高其肠道消化酶活性，促进营养物质的吸收。胞质蛋白闭锁小带蛋白（zonula occludens，ZO）和跨膜蛋白occludin、claudins组成的紧密连接蛋白（tight junctions，TJs）是保持细胞间结构完整性的重要蛋白[86]。在体外实验中，异亮氨酸通过预防和修复途径可以抑制Cu诱导的建鲤原代肠上皮细胞的氧化损伤，从而保护肠上皮细胞的结构完整性[82]。在建鲤[21]的研究表明，异亮氨酸的缺乏上调了肠道 occludin、claudin-3 和 claudin-7 mRNA表达水平。这个结果与草鱼的研究不一致，研究表明适宜的异亮氨酸可以提高Cu应激后草鱼鳃中 claudin-3、claudin-b、claudin-c、ZO-1 和 occludin mRNA表达水平[17]，从而增强细胞间的紧密连接。为了进一步阐明异亮氨酸与紧密蛋白的关系，Zhao等[82]检测了调节紧密连接蛋白的MAPKs信号通路，发现异亮氨酸对TJs基因表达的调控可能与下调蛋白激酶（extracellular regulated protein kinases，ERK1）、提高p38蛋白激酶（p38 mitogen activated protein kinase，p38MAPK）基因表达有关。关于异亮氨酸对紧密连接蛋白相关基因的调控机制还需要深入的研究。

肠道微生物群对宿主动物的肠道健康起着重要作用。在建鲤[21]的研究中发现，异亮氨酸促进了乳酸菌（Lactobacillus）和芽孢杆菌（Bacillus）的生长，抑制了嗜水气单胞菌（Aeromonas hydrophila）和大肠杆菌（Escherichia coli）的生长。乳酸杆菌在抑制肠道内有害菌生长的同时，对TJs蛋白的分布与表达起到促进作用[87]，从而对肠道起到保护作用。而嗜水气单胞菌和大肠杆菌则会破坏动物的肠道菌群结构，这说明异亮氨酸可以通过改善菌群结构促进鱼类肠道健康。

五、异亮氨酸对肉质的影响

异亮氨酸作为支链氨基酸，不同于非支链氨基酸，其主要在肌肉中分解代谢产生谷氨酸[66]。膳食中补充谷氨酸可以增强大西洋鲑鱼片的硬度[88]。基于此，Gan 等[11]研究了异亮氨酸对草鱼肌肉品质的影响。肌肉硬度和保水率是鱼肉质量的两个重要参数，这些参数的下降通常被认为是鱼片质量损失的一个指标[89]。研究表明，饲料中缺乏（3.8 g/kg）或添加过量（18.5 g/kg）的异亮氨酸会使草鱼幼鱼肌肉剪切力与 9.3 g/kg 异亮氨酸组相比下降 20% 和 30%，同时异亮氨酸缺乏和过量导致肌肉蒸煮损失与 12.5 g/kg 异亮氨酸组相比分别增加近 24% 和 26%。组织蛋白酶 B 和 L 在鱼类肌肉降解过程中起关键作用[90]。饲料中添加 9.3～12.5 g/kg 异亮氨酸可以降低因缺乏和过量的异亮氨酸引起的蛋白酶 B 和 L 活性的增强。Ofstad 等人[91]证明，胶原将肌肉纤维和纤维束粘在一起，有助于鱼类肌肉的持水。缺乏异亮氨酸会导致草鱼幼鱼肌肉乳酸浓度的升高，当给予足够水平的异亮氨酸时，可降低乳酸浓度。乳酸浓度与肌肉 pH 呈负相关[92]，而鱼的肌肉 pH 下降通常伴随着肌肉软化和水分流失[93]。这表明，异亮氨酸可能通过降低鱼体内乳酸浓度，提高了肌肉的 pH，从而适当改善鱼类的肉质。

第五节　小结与展望

目前，研究人员已经开展了大量关于水产动物对异亮氨酸的需要量的研究，鱼类对异亮氨酸的需要量为 0.47%～1.98%，虾蟹类对异亮氨酸的需要量为 1.01%～2.25%，而仿刺参对异亮氨酸的需要量为 0.98%。同时，主要研究了异亮氨酸与其他两个支链氨基酸之间的相互作用，开展了大量的关于异亮氨酸提高机体蛋白质代谢、免疫和抗氧化功能的研究。但是，关于异亮氨酸对肠道健康和肉质的影响的报道相对较少，尤其是对其调控机制值得深入的研究。此外，异亮氨酸还具有促进胰岛素分泌的作用，暗示其可以调节机体的糖脂代谢，未来有必要开展异亮氨酸对水产动物糖脂代谢影响的研究。

参考文献

[1] 邵继智. 支链氨基酸的生化与营养 [J]. 氨基酸杂志，1990，(2)：22-35.

[2] Collier R T, Mcnamara J P, Wallace C R, et al. A review of endocrine regulation of metabolism during lactation [J]. Animal Science, 1994, 59: 498.

[3] 李爱杰. 水产动物营养与饲料学 [M]. 北京：中国农业出版社，1996.

[4] 尚晓迪. 草鱼幼鱼对异亮氨酸需要量的研究 [D]. 重庆：西南大学，2009.

[5] 刘春生，张大鹏，刘文宽，等. 支链氨基酸在泌乳母猪营养中的研究现状 [J]. 饲料工业，2006 (1)：25-26.

[6] Ahmed I, Khan A. Dietary branched-chain amino acid valine, isoleucine and leucine requirements of fingerling Indian major carp, *Cirrhinus mrigala* (Hamilton) [J]. British Journal of Nutrition, 2006,

96 (3)：450-460.

[7] Khan M A, Abidi S F. Dietary isoleucine requirement of fingerling Indian major carp, *Labeo rohita* (Hamilton) [J]. Aquaculture Nutrition, 2007, 13 (6)：424-430.

[8] Wilson R P, Poe W E, Robinson E H. Leucine, isoleucine, valine and histidine requirements of fingerling channel catfish [J]. Journal of Nutrition, 1980 (4)：627-633.

[9] 杨霞，叶金云，周志金，等. 中华绒螯蟹幼蟹对亮氨酸和异亮氨酸的需要量 [J]. 水生生物学报, 2014, 38 (6)：1062-1070.

[10] 尚晓迪，罗莉，文华，等. 草鱼幼鱼对异亮氨酸的需要量 [J]. 水产学报, 2009, 33 (05)：813-822.

[11] Gan L, Jiang W D, Pei W, et al. Flesh quality loss in response to dietary isoleucine deficiency and excess in fish：A link to impaired Nrf2-dependent antioxidant defense in muscle [J]. Plos One, 2014, 9 (12)：115-129.

[12] Zehra S, Khan M A. Dietary isoleucine requirement of fingerling catla, *Catla catla* (Hamilton), based on growth, protein productive value, isoleucine retention efficiency and carcass composition [J]. Aquaculture International, 2013, 21 (6)：1243-1259.

[13] Ravi J, Devaraj K V. Quantitative essential amino acid requirements for growth of catla, *Catla catla* (Hamilton) [J]. Aquaculture, 1991, 96 (3-4)：281-291.

[14] Santiago C B, Lovell R T. Amino acid requirements for growth of Nile Tilapia [J]. Journal of Nutrition, 1988, 118：1540-1546.

[15] Fa A, Mm B, Mos C, et al. Assessment of isoleucine requirement of fast-growing Nile tilapia fingerlings based on growth performance, amino acid retention, and expression of muscle growth-related and mTOR genes [J]. Aquaculture, 2021, 539：1-8.

[16] 李燕. 鲈鱼和大黄鱼支链氨基酸与组氨酸营养生理的研究 [D]. 青岛：中国海洋大学, 2010.

[17] Feng L, Gan L, Jiang W D, et al. Gill structural integrity changes in fish deficient or excessive in dietary isoleucine：towards the modulation of tight junction protein, inflammation, apoptosis and antioxidant defense via NF-κB, TOR and Nrf2 signaling pathways [J]. Fish & Shellfish Immunology, 2017, 63：127-138.

[18] Ren M, Habte-tsion H M, Bo L, et al. Dietary isoleucine requirement of juvenile blunt snout bream, *Megalobrama amblycephala* [J]. Aquaculture Nutrition, 2017, 23 (2)：322-330.

[19] Zhao J, Liu Y, Jiang J, et al. Effects of dietary isoleucine on growth, the digestion and absorption capacity and gene expression in hepatopancreas and intestine of juvenile Jian carp (*Cyprinus carpio* var. Jian) [J]. Aquaculture, 2012, 368-369：117-128.

[20] Zhao J, Liu Y, Jiang J, et al. Effects of dietary isoleucine on the immune response, antioxidant status and gene expression in the head kidney of juvenile Jian carp (*Cyprinus carpio* var. Jian) [J]. Fish & Shellfish Immunology, 2013, 35：572-580.

[21] Zhao J, Feng L, Liu Y, et al. Effect of dietary isoleucine on the immunity, antioxidant status, tight junctions and microflora in the intestine of juvenile Jian carp (*Cyprinus carpio* var. Jian) [J]. Fish & Shellfish Immunology, 2014, 41 (2)：663-673.

[22] Benakappa S, Varghese T. Isoleucine, leucine, and valine requirement of juvenile Indian major carp, *Cirrhinus cirrhosus* (Bloch, 1795) [J]. Acta Ichthyol. Piscatoria, 2003, 33：161-172.

[23] 路凯，徐玮，麦康森，等. 生长中期花鲈对L-异亮氨酸需要量的研究 [J]. 水产学报, 2015, 39 (2)：203-212.

[24] Neu D, Boscolo W, Almeida F, et al. Growth performance, hematology, and muscle growth in

isoleucine fed Nile tilapia [J]. Boletim do Instituto de Pesca, 2017, 43: 231 - 242.

[25] 孙玉轩. 吉富罗非鱼亮氨酸和异亮氨酸需要量的研究 [D]. 重庆：西南大学, 2014.

[26] Borlongan I G, Coloso R M. Requirements of juvenile milkfish (*Chanos chanos Forsskal*) for essential amino acids [J]. Journal of Nutrition, 1993, 123 (1): 125 - 132.

[27] Sharf Y, Khan M A. Effect of dietary isoleucine level on growth, protein retention efficiency, haematological parameter, lysozyme activity and serum antioxidant status of fingerling *Channa punctatus* (Bloch) [J]. Aquaculture Nutrition, 2020, 26: 908 - 920.

[28] Castillo S, Gatlin Ⅲ D M. Dietary requirements for leucine, isoleucine and valine (branched - chain amino acids) by juvenile red drum *Sciaenops ocellatus* [J]. Aquaculture Nutrition, 2018, 24 (3): 1056 - 1065.

[29] 孙志远, 卫育良, 徐后国, 等. 红鳍东方鲀幼鱼异亮氨酸营养需求的研究 [J]. 动物营养学报, 2020, 32 (11): 5347 - 5356.

[30] Chance R E, Mertz E T, Halver J E. Nutrition of salmonoid fishes. Ⅶ. Isoleucine, leucine, valine and phenylalanine requirements of chinook salmon and inter - relations between isoleucine and leucine for growth [J]. Journal of Nutrition, 1964, 83 (21): 177 - 185.

[31] Zhou Z, Wu X, Li X, et al. The optimum dietary isoleucine requirement of juvenile hybrid grouper (*Epinephelus fuscoguttatus* ♀ × *Epinephelus lanceolatus* ♂) [J]. Aquaculture Nutrition, 2020, 26: 1295 - 1310.

[32] Rodehutscord M, Becker A, Pack M, et al. Response of rainbow trout (*Oncorhynchus mykiss*) to supplements of individual essential amino acids in a semipurified diet, including an estimate of the maintenance requirement for essential amino acids [J]. Journal of Nutrition, 1997, 127: 1166 - 1175.

[33] Steven G H, Gary L R, Malden C N. Dietary requirements for essential branched - chain amino acids by lake trout [J]. Transactions of the American Fisheries Society, 1983, 112 (6): 812 - 817.

[34] Huang Z, Lin H Z, Peng J S, et al. Effects of dietary isoleucine levels on the growth performance, feed utilization, and serum biochemical indices of juvenile golden pompano, *Trachinotus ovatus* [J]. Israeli Jounal of Aquaculture Bamidgeh, 2015, 1226: 1 - 9.

[35] Millamena O M, Teruel M B, Kanazawa A, et al. Quantitative dietary requirements of post larval tiger shrimp, *Penaeus monodon*, for histidine, isoleucine, leucine, phenylalanine and tryptophan [J]. Aquaculture, 1999, 179 (1): 169 - 179.

[36] Liu F J, Liu Y J, Tian L X, et al. Quantitative dietary isoleucine requirement of juvenile Pacific white shrimp, *Litopenaeus vannamei* (Boone) reared in low - salinity water [J]. Aquaculture International, 2014, 22: 1481 - 1497.

[37] 韩秀杰. 仿刺参 (*Apostichopus japonicus*) 幼参对蛋氨酸、缬氨酸和异亮氨酸最适需要量的研究 [D]. 上海：上海海洋大学, 2018.

[38] Kaushik S J, Seiliez I. Protein and amino acid nutrition and metabolism in fish: current knowledge and future needs [J]. Aquaculture Research, 2010, 41: 322 - 332.

[39] Mai K, Xue M, He G, et al. Chapter 4 - protein and amino acid [C]//Haedy R, Kaushik S. Fish nutrition. Academic press, 2021, 181 - 302.

[40] Hua K, Bureau D. Estimating changes in essential amino acid requirements of rainbow trout and Atlantic salmon as a function of body weight or diet composition using a novel factorial requirement model [J]. Aquaculture, 2019, 513: 34440.

[41] Walton M J, Cowey C B, Adron J W. Methionine metabolism in rainbow trout fed diets of different

methionine and cystine content [J]. Journal of Nutrition, 1982, 112: 1525-1535.

[42] Kim K I, Kayes T B. Requirement for sulphur amion acids and utilization of D-methionine by rainbow trout [J]. Aquaculture, 1992, 101: 95-103.

[43] Do Nascimento T, Mansano C, Peres H, et al. Determination of the optimum dietary essential amino acid profile for growing phase of Nile tilapia by deletion method [J]. Aquaculture, 2020, 523: 735204.

[44] Monentcham S, Whatelet B, Pouomogne V, et al. Egg and whole-body amino acid profile of African bonytongue (*Heterotis niloticus*) with an estimation of their dietary indispensable amino acids requirements [J]. Fish Physiology and Biochemistry, 2010, 36: 531-538.

[45] Gurure R, Atkinson J, Moccla R. Amino acid composition of Arctic charr, *Salvelinus alpinus* (L.) and the prediction of dietary requirements for essential amino acids [J]. Aquaculture Nutrition, 2007, 13: 266-272.

[46] De Almeida B A, Cyrino J. Estimating amino acid requirement of Brazilian freshwater fish from muscle amino acid profile [J]. Journal of the Word Aquaculture Society, 2009, 40: 818-823.

[47] Furuya W, Michelato M, Salaro Al, et al. Estimation of the dietary essential amino acid requirements of colliroja *Astyanax fasciatus* by using the ideal protein concept [J]. Latin American Journal of Aquatic Research, 2015, 43: 888-894.

[48] Ogino C. Requirements of carp and rainbow trout for essential amino acids [J]. Bulletin Japan Society Science Fish, 1980, 46: 171-174.

[49] Kaushik S. Whole body amino acid composition of European seabass (*Dicentrarchus labrax*), gilthead seabream (*Sparus aurata*) and turbot (*Psetta maxima*) with an estimation of their IAA requirement profiles [J]. Aquatic Living Resources, 1998, 11: 355-358.

[50] Gaber M, Salem M, Zaki M, et al. Amino acid requirements of gilthead bream (*Sparus aurata*) juveniles [J]. World Journal of Engineering and Technology, 2016, 4: 18-24.

[51] Peres H, Oliva-teles A. The optimum dietary essential amino acid profile for gilthead seabream (*Sparus aurata*) juveniles [J]. Aquaculture, 2009, 296: 81-86.

[52] Tesser M, Silva E, Sampaio L. Whole-body and muscle amino acid composition of plata pompano (*Trachinotus marginatus*) and prediction of dietary essential amino acid requirements [J]. Revista Colombiana de Ciencias Pecuarias, 2014, 27: 299-305.

[53] Forster I, Ogata H. Lysine requirement of juvenile Japanese flounder *Paralichthys olivaceus* and juvenile red sea bream *Pagrus major* [J]. Aquaculture, 1998, 161: 131-142.

[54] Ogunji J, Wirth M, Rennert B. Assessing the dietary amino acid requirements of tilapia, *Oreochromis niloticus* fingerlings [J]. Journal Aquaculture in the Tropics, 2005, 20: 241-250.

[55] Diógenes A, Fernandes J, Dorigam J, et al. Establishing the optimal essential amino acid ratios in juveniles of Nile tilapia (*Oreochromis niloticus*) by the deletion method [J]. Aquaculture Nutrition, 2016, 22: 435-443.

[56] Khan K, Mansano C, Nascimento T, et al. Whole-body amino acid pattern of juvenile, preadult, and adult pacu, *Piaractus mesopotamicus*, with an estimation of its dietary essential amino acid requirements [J]. Journal of the World Aquaculture Society, 2019, 51: 224-234.

[57] Kasozi N, Iwe G, Sadik K, et al. Dietary amino acid requirements of pebbly fish, *Alestes baremoze* (Joannis, 1835) based on whole body amino acid composition [J]. Aquaculture Reports, 2019, 14: 100197.

[58] Moon H, Gatlin III D. Total sulfur amino acid requirement of juvenile red drum, *Sciaenops ocellatus*

[J]. Aquaculture, 1991, 95: 97-106.

[59] Hossain M, Almatar S, James C. Whole body and egg amino acid composition of silver pomfret, *Pampus argenteus* (Euphrasen, 1788) and prediction of dietary requirements for essential amino acids [J]. Journal of Applied Ichthyology, 2011, 27: 1067-1071.

[60] Small B, Soares J. Estimating the quantitative essential amino acid requirements of striped bass *Morone saxatilis*, using fillet A/E ratios [J]. Aquaculture Nutrition, 1998, 4: 225-232.

[61] Ng W, Hung S. Estimating the ideal dietary indispensable amino acid pattern for growth of white sturgeon, *Acipenser transmontanus* (Richardson) [J]. Aquaculture Nutrition, 1995, 1: 85-94.

[62] Wang L, Han Y, Jiang Z, et al. Interactive effects of dietary leucine and isoleucine on growth, blood parameters, and amino acid profile of Japanese flounder *Paralichthys olivaceus* [J]. Fish Physiology and Biochemistry, 2017, 43: 1265-1278.

[63] Kajikawa H, Tajima K, Mitsumori M, et al. Inhibitory effects of isoleucine and antagonism of the other branched-chain amino acids on fermentation parameters by mixed ruminal microbes in batch cultures and rumen simulating fermenters (Rusitec) [J]. Animal Science Journal, 2007, 78: 266-274.

[64] Han Y, Han R, Koshio S, et al. Interactive effects of dietary valine and leucine on two sizes of Japanese flounder *Paralichthys olivaceus* [J]. Aquaculture, 2014, 432: 130-138.

[65] Shi L, Zhai H, Wei L, et al. Interactive effect between dietary valine and isoleucine on growth performance, blood parameters and resistance against low salinity stress of Japanese flounder *Paralichthys olivaceus* [J]. Journal of Applied Ichthyology, 2021, 37: 285-294.

[66] 王莉苹. 牙鲆（*Paralichthys olivaceus*）饲料中亮氨酸与异亮氨酸交互作用的研究 [D]. 大连：大连海洋大学, 2017.

[67] Choo P-S, Smith T K, Cho C Y, et al. Dietary excess of leucine influence growth and body composition of rainbow trout [J]. Journal of Nutrition, 1991, 121: 1932-1939.

[68] Robinson E H, Poe W E, Wilson R P. Effects of feeding diets containing an imbalance of branched chain amino acids on fingerling channel catfish [J]. Aquaculture, 1984, 37: 51-62.

[69] Yaghoubi M, Mozanzadeh M T, Marammazi J G, et al. Effects of dietary essential amino acid deficiencies on the growth performance and humoral immune response in silvery-black porgy (*Sparidentex hasta*) juveniles [J]. Aquaculture Research, 2017, 48: 5311-5323.

[70] Doi M, Yamaok I, Nakayama M, et al. Isoleucine, a blood glucose-lowering amino acid, increases glucose uptake in rat skeletal muscle in the absence of increases in AMP-activated protein kinase activity [J]. Journal of Nutrition, 2005, 135: 2103-2108.

[71] Johnston I A, Bower N I, Macqueen D J. Growth and the regulation of myotomal muscle mass in teleost fish [J]. Journal of Experimental Biology, 2011, 214: 1617-1628.

[72] Avruch J, Hara K, Lin Y, et al. Insulin and amino-acid regulation of mTOR signaling and kinase activity through the Rheb GTPase [J]. Oncogene, 2006, 25 (48): 6361-6372.

[73] Kawanago M, Takemura S, Ishizuka R, et al. Dietary branched-chain amino acid supplementation affects growth and hepatic insulin-like growth factor gene expression in yellowtail, *Seriola quinqueradiata* [J]. Aquaculture Nutrition, 2015, 21: 63-72.

[74] 甘露. 异亮氨酸对生长中期草鱼肉质和鳃屏障功能的影响 [D]. 成都：四川农业大学, 2014.

[75] 黄康, 梁旭方, 何珊, 等. 异亮氨酸对鳜 mTOR 信号通路及氮代谢影响 [J]. 水生生物学报, 2018, 42 (05): 879-886.

[76] Jobling M. Some effects of temperature, feeding and body weight on nitrogenous excretion in young

plaice *Pleuronectes platessa* L. [J]. Journal of Fish Biology, 1981, 18 (1): 87-96.

[77] 银龙, 周小秋, 赵娟, 等. 异亮氨酸对鱼类非特异性免疫功能影响的研究进展 [J]. 动物营养学报, 2019, 31 (07): 3023-3030.

[78] Rahimnejad S, Lee K J. Isoleucine effects on non-specific immune response of juvenile olive flounder (*Paralichthys olivaceus*) [J]. Fish and Shellfish Immunology, 2013, 34 (6): 1731-1732.

[79] 黄忠, 周传鹏, 林黑着, 等. 饲料异亮氨酸水平对卵形鲳鲹消化酶活性和免疫指标的影响 [J]. 南方水产科学, 2017, 13 (1): 50-57.

[80] Weichha R T T, Costantino G, Poglitschm, et al. The TSC-mTOR signaling pathway regulates the innate inflammatory response [J]. Immunity, 2008, 29 (4): 565-577.

[81] Ryu J M, Lee M Y, Yun S P, et al. High glucose regulates cyclin D1/E of human mesenchymal stem cells through TGF-β1 expression via Ca^{2+}/PKC/M APKs and PI3K/Akt/mTOR signal pathways [J]. Journal of Cellular Physiology, 2010, 224 (1): 59-70.

[82] Zhao J, Wu P, Jiang W, et al. Preventive and reparative effects of isoleucine against copper-induced oxidative damage in primary fish enterocytes [J]. Fish Physiology and Biochemistry, 2017, 43 (4): 1021-1032.

[83] 王莉苹, 姜志强, 孙梦蕾, 等. 饲料中亮氨酸和异亮氨酸交互作用对牙鲆消化酶及免疫相关酶活力的影响 [J]. 大连海洋大学学报, 2017, 32 (03): 287-293.

[84] Ma Q. Role of nrf2 in oxidative stress and toxicity [J]. Annual Review Pharmacology and Toxicology, 2013, 53: 401-426.

[85] 王旭, 周婧, 薛晓强, 等. 牙鲆饲料中异亮氨酸与缬氨酸的交互作用对消化酶和部分免疫酶的影响 [J]. 饲料工业, 2018, 39 (20): 23-28.

[86] Niklasson L. Intestinal mucosal immunology of salmonids response to stress and infection and crosstalk with the physical barrier [D]. Geoteborg: University of Gothenburg, 2013.

[87] Karczewski J, Troost F J, Konings I, et al. Regulation of human epithelial tight junction proteins by *Lactobacillus plantarum in vivo* and protective effects on the epithelial barrier [J]. American Journal of Physiology: Gastrointestinal and Liver Physiology, 2010, 298 (6): G851-G859.

[88] Larsson T, Koppang EO, Espe M, et al. Fillet quality and health of Atlantic salmon (*Salmo salar* L.) fed a diet supplemented with glutamate [J]. Aquaculture, 2014, 426: 288-295.

[89] Liu D, Liang L, Xia W, et al. Biochemical and physical changes of grass carp (*Ctenopharyngodon idella*) fillets stored at 23 and 0 ℃ [J]. Food Chemistry, 2013, 140: 105-114.

[90] Bahuaud D, Morkore T, Østbye Tk, et al. Muscle structure responses and lysosomal cathepsins B and L in farmed Atlantic salmon (*Salmo salar* L.) pre- and postrigor fillets exposed to short and long-term crowding stress [J]. Food Chemistry, 2010, 118: 602-615.

[91] Ofstad R, Kidman S, Myklebust R, et al. Liquid holding capacity and structural changes during heating of fish muscle: cod (*Gadus morhua* L.) and salmon (*Salmo salar*) [J]. Food structure, 1993, 12: 163-174.

[92] Acerete L, Reig L, Alvarez D, et al. Comparison of two stunning/slaughtering methods on stress response and quality indicators of European sea bass (*Dicentrarchus labrax*) [J]. Aquaculture, 2009, 287: 139-144.

[93] Ayala M D, Santaella M, Marti'Nez C, et al. Muscle tissue structure and flesh texture in gilthead sea bream, *Sparus aurata* L., fillets preserved by refrigeration and by vacuum packaging [J]. LWT-Food science and Technology, 2011, 44: 1098-1106.

第九章 缬 氨 酸

支链氨基酸在鱼类生长中具有调节蛋白质合成与降解、氧化供能及提高免疫反应的重要作用[1]。缬氨酸、亮氨酸和异亮氨酸同为支链氨基酸,是鱼类10种必需氨基酸之一,主要在全身肌肉组织中进行氧化代谢,小部分与其他氨基酸相同在肝脏代谢[2]。缬氨酸为生糖氨基酸,经缬氨酸转氨酶脱掉氨基后形成α-酮异戊酸,经缬氨酸脱氢酶复合物催化氧化脱羧生成脂酰CoA,脂酰CoA通过β氧化作用经β羧酰基CoA的中间过程,最终成为琥珀酰CoA参与三羧酸循环[3]。

缬氨酸对不同种类水产动物均具有营养调控作用。缬氨酸是人工及野生大西洋鲑胚胎和幼鱼发育不可缺少的氨基酸[4],也是中华绒螯蟹生长的第一限制氨基酸[5]。添加1.50%的包膜缬氨酸提高了凡纳滨对虾虾苗的成活率,缓解了虾苗开口摄食后第9天生长速度的下降[6]。与其他支链氨基酸促进采食的机制不同,脑内注射缬氨酸不仅能够抑制鳜下丘脑的神经元中促食基因 *AGRP*(agouti-related peptide)和 *NPY*(neuropeptide Y)的表达,同时抑制神经元中雷帕霉素靶蛋白(mTOR)信号通路的表达从而联合抑制采食[7]。本章综述了水产动物缬氨酸的需要量、缬氨酸与其他支链氨基酸的相互作用、缬氨酸对蛋白质代谢、免疫性能、抗氧化功能和肠道健康的影响,以期为缬氨酸在水产动物营养需要、功能和作用机理与健康养殖方面的深入研究提供参考。

第一节 缬氨酸的理化性质

缬氨酸(Valine,Val),分子式 $C_5H_{11}NO_2$,分子质量117.15,分子结构如图9-1所示,熔点为315℃,为白色结晶或结晶性粉末,溶于水,几乎不溶于乙醇。

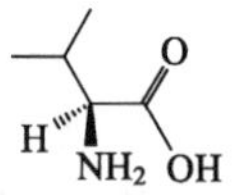

图9-1 缬氨酸分子结构

第二节 水产动物缬氨酸需要量

缬氨酸是水产动物的必需氨基酸。缬氨酸缺乏显著降低20 g及625 g尼罗罗非鱼的体重、体增重、采食量和饲料效率[8,9]。饲料缬氨酸的缺乏通过抑制以肝脏肉毒碱棕榈酰基转移酶-Ⅰ为代表的β-氧化基因表达量,以及肝脏固醇调节元件结合蛋白-1和肝脏脂肪酸合成酶等脂肪合成相关基因表达量,导致军曹鱼鱼体脂肪沉积量降低[10]。饲料中缺乏缬氨酸，显著降低 1.41 g 的大西洋鲑的增重率、饲料效率和蛋白质利用率[11]。过量缬氨酸对不同种类鱼的生长性能产生的作用不同。饲料中过量缬氨酸可降低卵形鲳鲹的存活率、生长性能、饲料利用率、体蛋白沉积率及消化酶活性[12]。过量的缬氨酸未对 3.15 g 红鼓鱼的生长性能产生显著影响[13]。另有研究表明，斑点叉尾鮰的饲料中过量添加缬氨酸也无法恢复前期缬氨酸缺乏导致的生长抑制和饲料效率的降低[14]。缺乏和过量缬氨酸对鱼类的生长性能、饲料效率、脂肪代谢、蛋白质代谢等产生不同程度的影响，因此，需要开展不同水产动物缬氨酸需要量的研究。

近年来发表的水产动物缬氨酸需要量见表 9-1。由表 9-1 可知，鱼类对缬氨酸的需要量为 0.70%～2.17%，甲壳类对缬氨酸的需要量为 1.08%～1.79%，仿刺参对缬氨酸的需要量为 1.79%。不同品种鱼对缬氨酸的需要量差异较大，草鱼需要量为 1.56%，吉富罗非鱼需要量为 1.41%，鲈需要量为 2.17%，表明肉食性鱼类缬氨酸需要量高于草食性和杂食性鱼类。同一品种不同规格的鱼对缬氨酸的需要量不同，不同规格草鱼[15,16,17]和鲈[1,18,19]缬氨酸需要量近似；但大规格尼罗罗非鱼对缬氨酸的需要量为 0.78%，低于小规格需要量 1.15%～1.40%[20,21,22]，大规格虹鳟对缬氨酸的需要量为 1.41%～1.50%，低于小规格需要量 1.57%[23,24]。缬氨酸需要量主要有占饲料比例和占饲料蛋白质比例两种方式，表 9-1 采用直接引用或折算为占饲料比例的方式进行需要量的表示。同一模型下不同评价指标对鱼类缬氨酸的需要量存在影响，研究人员通过多项评价指标，如同时计算增重率、蛋白质累积率及饲料效率等，综合评估水产动物缬氨酸需要量，这种方式是应用较为普遍的需要量评价策略[25,26]。此外，二次多项式模型预测异育银鲫的缬氨酸需要量为 1.72%，而折线模型为 1.21%，说明需要量受不同统计模型的影响[27]，折线模型一般会比其他模型（指数、曲线模型等）估算的需要量低[28]。在体重、投喂频率、实验时长和循环水环境条件基本相同的情况下，6.5 g 尼罗罗非鱼的缬氨酸需要量高于 6.48 g 的需要量，这可能是由于饲养温度（分别为 25～27 ℃和 27～29 ℃）及饲喂量［占体重（6%～8%）/d 和（8%～10%）/d］的不同产生了差异[20,21]。综上所述，水产动物缬氨酸需要量的差异可能受种类、规格、饲料组成、投喂次数及水平、饲养环境条件等多种因素的影响。

由于不同物种之间有一定程度的同质性[29]，因此将鱼类和甲壳类水产动物的缬氨酸需要量共同进行 meta 分析。基于不同研究的品种、初始大小、生长速率以及实验条件等的差异，在本章引用的参考文献中，将体增重率（WGR）进行标准化响应处理后与需要量进行分析（图 9-2）。为了分析剂量效应，使用了四参数营养动力学分析方法[30]。根据不同品种对缬氨酸的需要量的研究数据，分别采用了 WGR 与缬氨酸占饲料百分比（图 9-2A）和占饲料中粗蛋白百分比（图 9-2B）绘制曲线，并尝试采用不同回归分析对曲线进行拟合，最终确定了三次多项式回归分析模型（拟合曲线公式分别为：$y_{饲料干物质}=10.73x^3-70.38x^2+147.34x-0.225\,2$，$R^2=0.985\,8$；$y_{饲料蛋白}=0.566\,8x^3-10.13x^2+55.866x+1.481$，$R^2=0.979\,8$），拟合得到鱼类缬氨酸需要量为 1.253（占饲料）和 3.189（占粗蛋白），甲壳类缬氨酸需要量 1.736（占饲料）和 4.334（占粗蛋白）。

表 9-1 水产动物缬氨酸需要量

品种	体重（g）	模型	指标	需要量（占饲料干物质,%）	参考文献
鱼类					
草鱼	1.35	二次多项式模型	增重率	1.368	孙丽慧等[15]
			蛋白质累积率	1.279	
			饲料效率	1.306	
草鱼	9.5	二次多项式模型	增重率	1.561	罗莉等[16]
			肌肉 RNA/DNA 比率	1.600	
			蛋白质沉积率	1.510	
草鱼	268.00	二次多项式模型	增重率	1.40	Luo 等[17]
吉富罗非鱼	37.82	二次多项式模型	增重率	1.43	武文一[3]
			饲料系数	1.41	
			蛋白质效率	1.49	
鲈	183.28	二次多项式模型	特定生长率	2.17	窦秀丽[1]
			饲料效率	2.14	
花鲈	182.33	二次多项式模型	特定生长率	2.17	卫育良[18]
			饲料效率	2.14	
鲈	8	二次多项式模型	特定生长率	2.11	李燕[19]
			饲料效率		
尼罗罗非鱼	6.5	二次多项式模型	增重率	1.39	韦庆杰[20]
			蛋白质沉积率	1.41	
尼罗罗非鱼	6.48	折线模型	增重率	1.15	Xiao 等[21]
			蛋白质沉积率	1.27	
尼罗罗非鱼	50	折线模型	增重率	0.78	Santiago 等[22]
虹鳟	49	最小二乘法模型	蛋白质沉积率	1.57	Rodehutscord 等[23]
虹鳟	498	折线模型	吸收后游离缬氨酸浓度	1.50	Bae 等[24]
			餐后血浆氨浓度	1.41	
			餐后血浆游离缬氨酸浓度	1.47	
印度鲮	0.63	二次多项式模型	增重	1.59	Ahmed 等[25]
			饲料效率	1.50	
			蛋白质沉积率	1.48	
印度鲮	0.16	折线模型	饲料效率	1.5	Abidi[26]
			蛋白质沉积率		
异育银鲫	3.17	二次多项式模型	特定生长率	1.72	李桂梅[27]
		折线模型		1.21	
红鲷	32.04	折线模型	增重率	0.90	Rahimnejad 等[31]
			蛋白质沉积率	0.99	

(续)

品种	体重（g）	模型	指标	需要量（占饲料干物质,%）	参考文献
团头鲂	10.2	折线模型	特定生长率	1.32	Ren 等[32]
		折线模型	饲料效率	1.26	
卡特拉鲃	1.2	折线模型	平均体增重	1.42	Ravi 等[33]
建鲤	9.67	二次多项式模型	特定生长率	1.37	董敏[34]
卵形鲳鲹	5.34	二次多项式模型	增重率	2.017	Huang 等[12]
			蛋白质沉积率	1.987	
红鼓鱼	2.83	折线模型	增重率	1.24	Castillo 等[35]
斑点叉尾鲖	201	折线模型	增重率	0.71	Wilson 等[36]
			饲料效率	0.70	
遮目鱼	0.35	折线模型	增重	0.89	Borlongan 等[37]
			饲料效率		
甲壳类					
凡纳滨对虾	0.31	折线模型	生长率	1.79	王用黎等[38]
			增重率		
虎虾	0.014	折线模型	增重率	1.35	Millamena 等[39]
对虾	0.79	—	全身必需氨基酸分析和蛋白质分析	1.08	Teshima[40]
海参					
仿刺参	9.20	折线模型	增重率	1.79	韩秀杰[41]

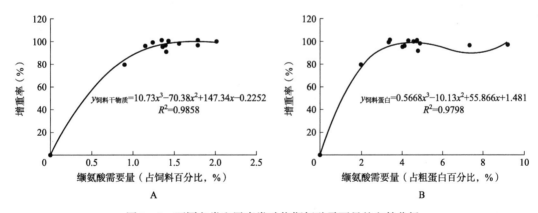

图9-2 不同鱼类和甲壳类动物缬氨酸需要量的文献分析

第三节 缬氨酸与支链氨基酸的相互作用

缬氨酸与支链氨基酸相互作用的研究，主要集中在其与亮氨酸关系的研究上。低亮氨酸饲料中添加缬氨酸能够缓解日本牙鲆的生长抑制，降低氧化应激反应[42]。添加需要量

2倍的亮氨酸，也无法弥补斑点叉尾鮰因缬氨酸缺乏导致的饲料效率降低的问题；相反的，缬氨酸能够缓解亮氨酸缺乏引发的生长抑制。另外，添加缬氨酸可以缓解过量亮氨酸和低缬氨酸饲料产生的生长抑制，并且提高饲料效率[14]。饲料中添加缬氨酸和异亮氨酸可缓解由过量亮氨酸导致的虹鳟生长性能和代谢的负面影响[43]。综上所述，缬氨酸能够调节饲料中缺乏或过量亮氨酸导致的鱼类的生长抑制，反之亮氨酸对缬氨酸缺乏造成的生长抑制不具备调节作用。

另外，饲料中2%异亮氨酸与2.27%缬氨酸的交互作用通过阻断肝脏中谷氨酸脱氢酶的活性和基因表达量降低脱氨基作用，同时抑制肌肉中的腺苷单磷酸脱氨酶的活性和基因表达量降低肌肉中氨的产生，从而降低3.57 g牙鲆氨氮的排泄[44]。

第四节　缬氨酸对水产动物蛋白质代谢的影响

缬氨酸可以促进鱼类蛋白质代谢，缺乏或过量缬氨酸均对蛋白质的沉积率和鱼体蛋白质含量产生不良影响。随着饲料中缬氨酸含量的增加，1.35 g和9.5 g草鱼的蛋白质沉积率显著增加，分别在缬氨酸水平为1.26%和1.63%时达到最大值[15,16]；268 g的草鱼饲料含有1.31%的缬氨酸时，肌肉中蛋白质含量最高[45]。含有1.317%～1.781%缬氨酸的饲料能够显著提高尼罗罗非鱼幼鱼的体蛋白效率和蛋白质沉积率[20]。饲料中缬氨酸含量达到1.04%～1.40%时，显著提高了红鼓鱼蛋白效率和蛋白质沉积率[35]。研究表明，支链氨基酸具有促进多肽链合成的起始来促进蛋白质的合成，并抑制其分解的作用[46]。脑内注射的缬氨酸通过抑制肝脏中谷氨酸脱氢酶和肌肉中腺苷酸脱氨酶活性、激活脑-肝回路的肝脏中mTOR复合物1的活性来减弱和阻断鳜外周器官的蛋白质的水解[47]。表9-1中显示，对氨基酸需要量的评价不仅局限于增重率、生长率和饲料效率等一般氨基酸的评价指标，蛋白质沉积率被大量使用，体现了缬氨酸对蛋白质代谢具有重要作用。上述研究表明，缬氨酸通过增加鱼类肌肉中蛋白质含量、提高蛋白质沉积率和抑制蛋白质分解的途径，综合调控蛋白质的代谢。

第五节　缬氨酸对水产动物免疫功能的影响

缬氨酸能够促进水产动物免疫器官的发育，以血清、肝脏和肠道为主要研究对象，提高水产动物的特异性免疫和非特异性免疫的功能。饲料中适宜含量的缬氨酸（0.87%～1.49%）不同程度地提高了幼建鲤头肾、后肾和脾脏的重量，促进幼建鲤免疫器官的生长发育；同时，提高白细胞吞噬率、总铁结合力、溶菌酶、酸性磷酸酶（ACP）活力、补体3（C3）和补体4（C4）的含量来增强其非特异性免疫能力，提高抗嗜水气单胞菌抗体效价和水平来增强其特异性免疫能力[34]。饲料中适宜含量的缬氨酸（1.66%～2.07%）与过高或过低缬氨酸含量相比，提高卵形鲳鲹血清中IgM、C3、C4和溶菌酶等免疫指标的含量[12]。缬氨酸通过提高尼罗罗非鱼肝脏溶菌酶的活性，增强肝脏免疫水平[21]。饲料中过高或过低的缬氨酸能够通过降低肠道溶菌酶、酸性磷酸酶活性及补体C3含量，提高促炎因子［白介素-8（IL-8）和肿瘤坏死因子α（TNF-α）］以及降低抑炎因子［白介素-18（IL-18）和转化生长因子β（TGF-β）］基因表达量抑制草鱼肠道免疫状态；当饲料中含有适量的缬氨酸（0.8%～1.67%）时，抑制核因子 κBP65/核因子抑制蛋白α/雷帕霉素靶点（nuclear factor kappa-B，NF-κB；inhibitor αof NF-κB P65，IκBα P65；target of rapamycin，TOR）信号通路基因的表达，进而抑制其下游促炎因子（IL-8、TNF-α）的活性，从而提高肠道免疫状态[17]。在缬氨酸含量达到1.06%～1.31%时，草鱼鳃中 IL-1β、IL-8、TNF-α 和 NF-κB P65 的表达量显著下降，IL-10、TGF-β 和 IκBα 的表达量显著升高，缓解过高或过低缬氨酸对草鱼鳃的免疫抑制作用[48]。上述研究表明，饲料中过高或过低的缬氨酸引发机体的免疫抑制，补充适量的缬氨酸可以缓解因缺乏缬氨酸导致的免疫抑制作用，并且提高动物体的免疫能力。

第六节 缬氨酸对水产动物抗氧化能力的影响

营养物质对动物抗氧化能力影响的评估指标，以总抗氧化能力（T-AOC）、过氧化物酶体标志物［过氧化氢酶（CAT）］、抗氧化酶体系［超氧化物歧化酶（SOD）、谷胱甘肽过氧化物酶（GPx）和谷胱甘肽（GSH）等］，脂质氧化产物［丙二醛（MDA）］为主要代表。饲料中缺乏或过量缬氨酸可通过降低卵形鲳鲹血清中 T-AOC 和 SOD、增加 MDA 含量抑制血清抗氧化水平[12]。饲料添加0.87%～1.49%缬氨酸通过降低幼建鲤免疫器官（头肾、后肾和脾脏）中 MDA 含量、提高 SOD 和 CAT 活性增强机体抗氧化能力[34]。饲料添加缬氨酸（1.14%～2.64%）能够通过不同程度地提高仿刺参肠道中 T-AOC、总超氧化物歧化酶（total superoxide dismutase，T-SOD）和 CAT 的活力提高肠道抗氧化功能[38]。缬氨酸（0.8%～1.91%）通过提高草鱼肌肉中 SOD、GSH、CAT 和 GPx，降低 MDA 的含量及基因表达量，提高肌肉抗氧化水平[45]。低亮氨酸（1.6%）和高缬氨酸（2.5%）水平的日本牙鲆氧化应激强度较低，对氧化应激的耐受能力较强；饲喂低亮氨酸（1.6%）和低缬氨酸（1.2%）组的日本牙鲆氧化应激强度较高，对氧化应激的耐受能力较低。因此，高缬氨酸能缓解低亮氨酸（1.6%）饲料对日本牙鲆细胞免受氧化应激的影响[42]。在缬氨酸含量达到1.31%～1.67%时，草鱼鳃中 MDA 和蛋白质羰基含量显著下降，灭活超氧阴离子和抗羟基自由基含量显著提高；在缬氨酸含量达到1.06%～1.31%时，Cu/Zn SOD 和 CAT 的活性显著增加；0.8%以上时显著提高 GPx、GSH、谷胱甘肽-s-转移酶和谷胱甘肽还原酶的活性。综上所述，缬氨酸在水产动物的免疫器官、血清、肠道和肌肉中都能够发挥抗氧化作用，提高氧化应激的耐受力，并且降低氧化应激水平[48]。

第七节 缬氨酸对肠道健康的影响

水产动物的生长与肠道健康密不可分。肠道自身的肠道屏障功能和消化能力，决定动物对有害物质的阻隔效应及对营养物质的吸收程度，并影响肠道中菌群的种类和丰度。缬氨酸（0.8%～1.91%）通过提高草鱼肠道紧密连接蛋白的闭合蛋白 claudin 家族（claudin b、claudin c、claudin 3、claudin 12 和 claudin 15）、闭锁蛋白（occludin）和闭锁小带蛋白-1（ZO-1）的表达量来改善肠道的物理屏障[45]。饲料含有0.87%～1.49%缬氨酸显著提高了幼建鲤肠道胰蛋白酶、糜蛋白酶、脂肪酶和淀粉酶的活力，0.87%～1.87%缬氨酸能够不同程度提高前肠、中肠和后肠皱襞的高度，并且提高肠道中乳酸菌、降低大肠杆菌和嗜水气单胞菌的数量[34]。缬氨酸能够为肠道菌群提供氮源，随着缬氨酸水平的增加，在缬氨酸含量为1.49%时，提高建鲤肠道乳酸菌和芽孢杆菌的丰度、抑制大肠杆菌的生长[49]。综上所述，缬氨酸能从提高肠道防御和消化性能以及调节菌群的种类和丰度两个方向，协同促进肠道健康。

第八节　小结与展望

现有研究显示，鱼类对缬氨酸的需要量为0.70%～2.17%，贝壳类对缬氨酸的需要量为1.08%～1.79%，仿刺参对缬氨酸的需要量为1.79%，以目前研究数据为基础拟合得到鱼类缬氨酸需要量为1.253（占饲料百分比）和3.189（占粗蛋白百分比），甲壳类缬氨酸需要量1.736（占饲料百分比）和4.334（占粗蛋白百分比），且缬氨酸对水产动物不同组织器官的蛋白质代谢、免疫性能、抗氧化功能、肠道健康功能都具有良好的促进作用。缬氨酸对水产动物的影响还在持续研究中，其作用机制的深入挖掘、不同种类和规格鱼需要量的完善、与其他营养物质之间的互作关系以及新的调控功能的探索，都将成为未来缬氨酸研究的方向。

参考文献

[1] 窦秀丽. 鲈鱼（*Lateolabrax japonicus*）生长中后期缬氨酸、组氨酸和苏氨酸营养需求的研究［D］. 大连：大连海洋大学，2014.

[2] National research council. Nutrient requirements of fish and Shrimp［M］. Washington, DC, USA：National Academies Press，2011.

[3] 武文一. 吉富罗非鱼对饲料精氨酸、苯丙氨酸和缬氨酸的需要量研究［D］. 上海：上海海洋大学，2016.

[4] Srivastava R K, Brown J A, Shahidi F. Changes in the amino acid pool during embryonic development of cultured and wild Atlantic salmon (*Salmo salar*)［J］. Aquaculture, 1995, 131 (1-2)：115-124.

[5] Xie Q, Liu Y. Study on the nutritional quality of ecologically bred Chinese mitten crabs with different body weights［J］. Aquaculture Research, 2020, 51 (7)：2948-2961.

[6] 黄宝生. 虾片中添加缬氨酸对凡纳滨对虾仔虾生长、存活的影响［D］. 保定：河北大学，2010.

[7] Ke C, Zhen Z, Jiao L, et al. Different regulation of branched-chain amino acid on food intake by TOR signaling in Chinese perch (*Siniperca chuatsi*)［J］. Aquaculture, 2021, 530：735792.

[8] Diogenes A F, Fernandes J B K, Dorigam J C P, et al. Establishing the optimal essential amino acid ratios in juveniles of Nile tilapia (*Oreochromis niloticus*) by the deletion method［J］. Aquaculture Nutrition, 2016, 22：435-443.

[9] Rodrigues A T, Mansano C, Khan K U, et al. Ideal profile of essential amino acids for Nile tilapia (*Oreochromis niloticus*) in the finishing growth phase［J］. Aquaculture Research, 2020, 51：4724-4735.

[10] 王震，徐玮，麦康森，等．饲料缬氨酸水平对军曹鱼鱼体脂肪含量、血浆生化指标和肝脏脂肪代谢基因表达的影响［J］．水生生物学报，2016，40（4）：744-751.

[11] Rollin X，Mambrini M，Abboudi T，et al. The optimum dietary indispensable amino acid pattern for growing Atlantic salmon（*Salmo salar* L.）fry［J］．British Journal of Nutrition，2003，90（5）：865-876.

[12] Huang Z，Tan X H，Zhou C P，et al. Effect of dietary valine levels on the growth performance，feed utilization and immune function of juvenile golden pompano，*Trachinotus ovatus*［J］．Aquaculture Nutrition，2018，24（1）：74-82.

[13] Sergio C，Delbert M G. Imbalanced dietary levels of branched-chain amino acids affect growth performance and amino acid utilization of juvenile red drum *Sciaenops ocellatus*［J］．Aquaculture，2018，497：17-23.

[14] Wilson R P，Poe W E，Robinson E H. Effects of feeding diets containing an imbalance of branched-chain amino acids on fingerling channel catfish［J］．Aquaculture，1984，7（1）：51-62.

[15] 孙丽慧，陈建明，沈斌乾，等．草鱼鱼种对饲料中缬氨酸需要量的研究［J］．上海海洋大学学报，2017，26（6）：900-908.

[16] 罗莉，王亚哥，李芹，等．草鱼幼鱼对缬氨酸需要量的研究［J］．动物营养学报，2010，22（3）：616-624.

[17] Luo J B，Feng L，Jiang W D，et al. The impaired intestinal mucosal immune system by valine deficiency for young grass carp（*Ctenopharyngodon idella*）is associated with decreasing immune status and regulating tight junction proteins transcript abundance in the intestine［J］．Fish & Shellfish Immunology，2014，40（1）：197-207.

[18] 卫育良，窦秀丽，梁萌青，等．大规格花鲈对饲料中缬氨酸的需要量研究［J］．饲料工业，2021，42（8）：47-53.

[19] 李燕．鲈鱼和大黄鱼支链氨基酸与组氨酸营养生理的研究［D］．青岛：中国海洋大学，2010.

[20] 韦庆杰．尼罗罗非鱼幼鱼对缬氨酸需要量的研究［D］．南京：南京农业大学，2015.

[21] Xiao W，Li D Y，Zhu J L，et al. Dietary valine requirement of juvenile Nile tilapia，*Oreochromis niloticus*［J］．Aquaculture Nutrition，2018，24（22）：315-323.

[22] Santiago C B，Lovell R T. Amino acid requirements for growth of Nile tilapia［J］．The Journal of Nutrition，1988，118（12）：1540.

[23] Rodehutscord M，Becker A，Pack M，et al. Response of rainbow trout（*Oncorhynchus mykiss*）to supplements of individual essential amino acids in a semipurified diet，including an estimate of the maintenance requirement for essential amino acids［J］．The Journal of Nutrition，1997，127（6）：1166-1175.

[24] Bae J，Park G，Yun H，et al. The dietary valine requirement for rainbow trout，*Oncorhynchus mykiss*，can be estimated by plasma free valine and ammonia concentrations after dorsal aorta cannulation［J］．Journal of Applied Animal Research，2012，40（1）：73-79.

[25] Ahmed I，Khan M A. Dietary branched-chain amino acid valine，isoleucine and leucine requirements of fingerling Indian major carp，*Cirrhinus mrigala*（Hamilton）［J］．British Journal of Nutrition，2006，96（3）：450-460.

[26] Abidi S F，Khan M A. Dietary valine requirement of Indian major carp，*Labeo rohita*（Hamilton）

fry [J]. Journal of Applied Ichthyology, 2004, 20 (2): 118 – 122.

[27] 李桂梅,解绶启,雷武,等. 异育银鲫幼鱼对饲料中缬氨酸需要量的研究 [J]. 水生生物学报, 2010, 34 (06): 1157 – 1165.

[28] National Research Council. Nutrientrequirements of fish and shrimp [M]. The National Academies Press, 2011: 11 – 12.

[29] Kaushik S J, Seiliez I. Protein and amino acid nutrition and metabolism in fish: current knowledge and future needs [J]. Aquaculture Research, 2010, 41: 322 – 332.

[30] Mercer L P. The quantitative nutrient-response relationship [J]. Journal of Nutrition, 1982, 112: 560 – 566.

[31] Rahimnejad S, Lee K J. Dietary valine requirement of juvenile red sea bream *Pagrus major* [J]. Aquaculture, 2013, 416 – 417 (2): 212 – 218.

[32] Ren M C, Habte-Tsion H M, Liu B, et al. Dietary valine requirement of juvenile blunt snout bream (*Megalobrama amblycephala* Yih, 1955) [J]. Journal of Applied Ichthyology, 2015, 31 (6): 1 – 7.

[33] Ravi J, Devaraj K V. Quantitative essential amino acid requirements for growth of catla, *Catla catla* (Hamilton) [J]. Aquaculture, 1991, 96: 281 – 291.

[34] 董敏. 缬氨酸与幼建鲤消化吸收能力,免疫能力以及氧化能力之间的关系 [D]. 雅安:四川农业大学, 2011.

[35] Castillo S, Gatlin D M. Dietary requirements for leucine, isoleucine and valine (branched-chain amino acids) by juvenile red drum *Sciaenops ocellatus* [J]. Aquaculture Nutrition, 2018, 24: 1056 – 1065.

[36] Wilson R P, Poe W E, Robinson E H. Leucine, isoleucine, valine and histidine requirements of fingerling channel catfish [J]. The Journal of Nutrition, 1980, 110 (4): 627 – 633.

[37] Borlongan I G, Coloso R M. Requirements of juvenile milkfish (*Chanos chanos* Forsskal) for essential amino acids [J]. The Journal of Nutrition, 1993, 123 (1): 125 – 132.

[38] 王用黎. 凡纳滨对虾幼虾对苏氨酸、亮氨酸、色氨酸和缬氨酸需要量的研究 [D]. 湛江:广东海洋大学, 2013.

[39] Millamena O M, Bautista-Teruel M N, Kanazawa A. Valine requirement of postlarval tiger shrimp, *Penaeus monodon* Fabricius [J]. Aquaculture Nutrition, 1996, 2 (3): 129 – 132.

[40] Teshima S, Alam M S, Koshio S, et al. Assessment of requirement values for essential amino acids in the prawn, *Marsupenaeus japonicus* (Bate) [J]. Aquaculture Research, 2002, 33 (6): 395 – 402.

[41] 韩秀杰. 仿刺参 (*Apostichopus japonicus*) 幼参对蛋氨酸、缬氨酸和异亮氨酸最适需要量的研究 [D]. 上海:上海海洋大学, 2018.

[42] Han Y Z, Han R Z, Shunsuke K, et al. Interactive effects of dietary valine and leucine on two sizes of Japanese flounder *Paralichthys olivaceus* [J]. Aquaculture, 2014, 432: 130 – 138.

[43] Yamamoto T, Shima T, Furuita H. Antagonistic effects of branched-chain amino acids induced by excess protein-bound leucine in diets for rainbow trout (*Oncorhynchus mykiss*) [J]. Aquaculture, 2004, 232 (1 – 4): 539 – 550.

[44] 王旭,周婧,薛晓强,等. 牙鲆饲料中异亮氨酸与缬氨酸的交互作用对消化酶和部分免疫酶的影响 [J]. 饲料工业, 2018, 39 (20): 23 – 28.

[45] 罗建波. 缬氨酸对生长中期草鱼生长、肉质和肠道免疫及其相关基因表达的影响 [D]. 雅安:四川农业大学, 2014.

［46］代腊，邹晓庭. 缬氨酸在畜禽生产中的应用 ［J］. 饲料与畜牧，2012（12）：20-23.

［47］Wang J，Liang X F，He S，et al. Valine acts as a nutritional signal in brain to activate TORC1 and attenuate postprandial ammonia-N excretion in Chinese perch（*Siniperca chuatsi*）［J］. Fish Physiology and Biochemistry，2020，46：2015-2025.

［48］Feng L，Luo J B，Jiang W D，et al. Changes in barrier health status of the gill for grass carp（*Ctenopharyngodon idella*）during valine deficiency：Regulation of tight junction protein transcript，antioxidant status and apoptosis-related gene expression ［J］. Fish & Shellfish Immunology，2015，45（2）：239-249.

［49］Dong M，Feng L，Kuang S Y，et al. Growth，body composition，intestinal enzyme activities and microflora of juvenile Jian carp（*Cyprinus carpio* var. Jian）fed graded levels of dietary valine ［J］. Aquaculture Nutrition，2012，19：1-14.

第十章 色氨酸

色氨酸（Tryptophan，Trp）又称β-吲哚基丙氨酸，是水产动物的必需氨基酸。色氨酸是饲料蛋白质的重要组成成分，是高脂肉粉、骨粉的第一限制氨基酸，是玉米蛋白粉、血粉、禽下脚料及蚕豆的第二限制性氨基酸，是羽扇豆、羽毛粉的第三限制性氨基酸[1]。色氨酸缺乏导致虹鳟生长性能降低、脊柱侧凸、前凸及白内障，肝脏、肾脏中钙、镁、钠及钾的含量升高[2]。在色氨酸缺乏饲料中补充晶体色氨酸，可以提高虹鳟、非洲鲇、杂交条纹鲈、露斯塔野鲮及凡纳滨对虾的生长性能[3,4,5,6,7]。色氨酸是5-羟色胺的前体物质，5-羟色胺是一种重要神经递质可以调节水产动物的摄食及攻击行为[8,9,10,11]。本章综述了色氨酸代谢途径、水产动物色氨酸的需要量以及色氨酸对摄食、应激、免疫的影响，以期为色氨酸在水产动物中开展深入研究提供参考。

第一节 色氨酸的理化性质

色氨酸分子式$C_{11}H_{12}N_2O_2$，分子质量204.23，分子结构如图10-1所示，熔点为281～282℃，外观为白色或微黄色结晶或结晶性粉末，无臭，味微苦。水中微溶，在乙醇中极微溶解，在氯仿中不溶，在甲酸中易溶，在氢氧化钠试液或稀盐酸中溶解。

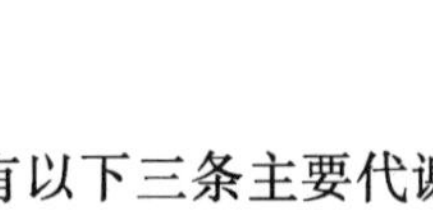

图10-1 色氨酸分子结构

第二节 色氨酸的代谢途径

饲料中的色氨酸被消化、吸收后，除用于蛋白质合成外，还有以下三条主要代谢途径（图10-2）。色氨酸在色氨酸羧化酶和芳香族氨基酸脱羧酶催化作用下合成5-羟色胺，进一步通过芳香烷基胺-N-乙酰基转移酶和羟基吲哚-氧-甲基转移酶催化合成褪黑素，此途径主要受生理状态和外部环境调节[12]。色氨酸主要通过犬尿氨酸-烟酸途径在肝脏中进行分解代谢生成N-甲酰犬尿氨酸，并受吲哚胺2,3-双加氧酶（IDO）和色氨酸2,3-双加氧酶（TDO）的调控，其中IDO是色氨酸代谢的限速酶[13]。N-甲酰犬尿氨酸在犬尿氨酸甲酰胺酶作用下生成L-犬尿氨酸，进一步生成3-羟基犬尿氨酸和3-羟基邻氨基苯甲酸，最后生成完全氧化的乙酰辅酶A，或者进一步生成喹啉酸和烟酸[14]。

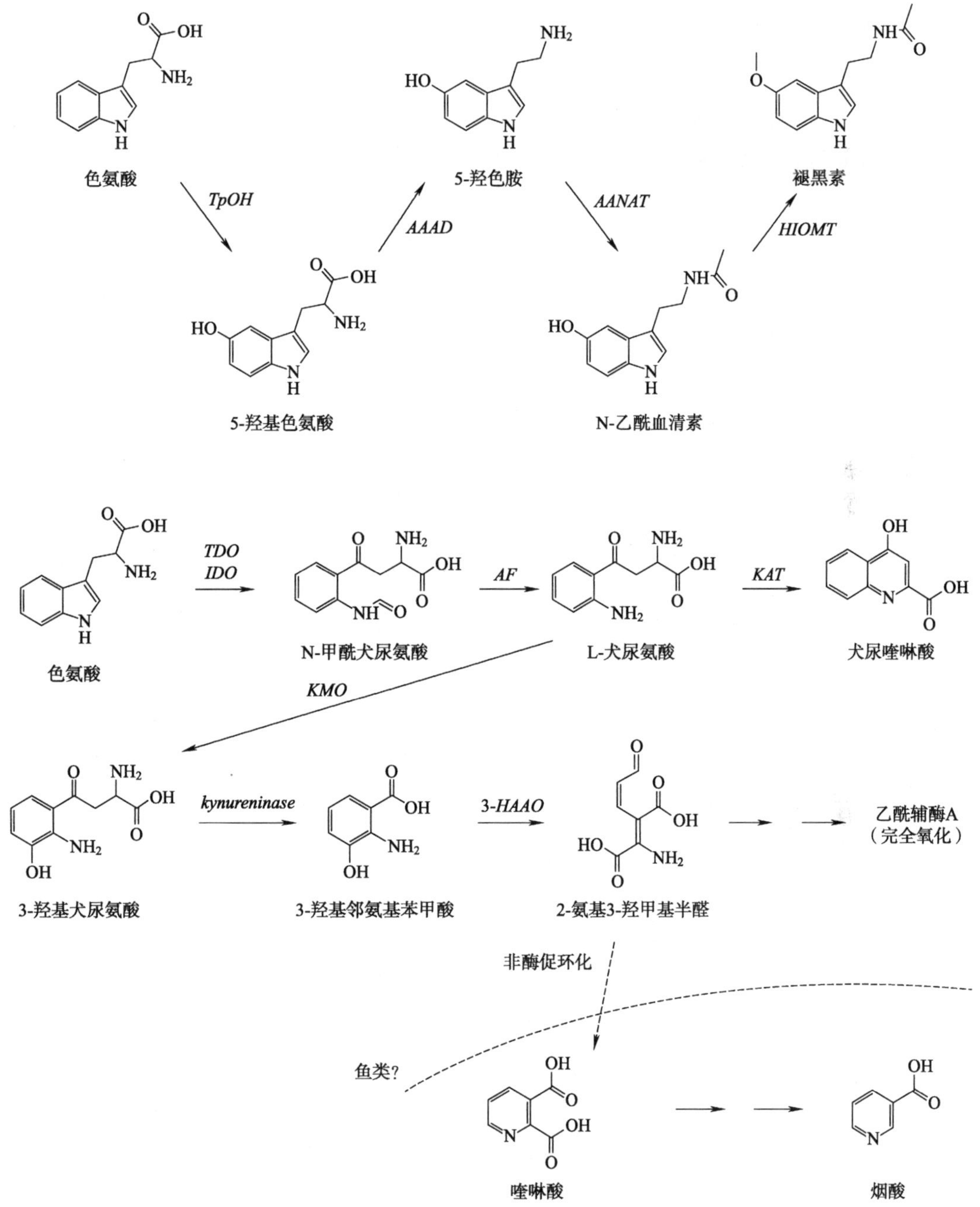

图 10-2 色氨酸代谢[15]

注：TpOH 表示色氨酸羧化酶，AAAD 表示芳香族氨基酸脱羧酶，AANAT 表示芳香烷基胺-N-乙酰基转移酶，HIOMT 表示羟基吲哚-氧-甲基转移酶，TDO 表示色氨酸 2,3-双加氧酶，IDO 表示吲哚胺 2,3-双加氧酶，AF 表示犬尿氨酸甲酰胺酶，KAT 表示犬尿氨酸氨基转移酶，KMO 表示犬尿氨酸 3-单加氧酶，3-HAAO 表示 3-羟基邻氨基苯甲酸 3,4-双氧化酶。

第三节 水产动物色氨酸需要量

色氨酸是水产动物机体蛋白质组成中含量最低的氨基酸。因此，水产动物饲料中色氨酸的需要量也低于其他必需氨基酸。水产动物色氨酸需要量与养殖品种、饲料组成、试验动物初始体重等多种因素有关，饲料中蛋白质、氨基酸及能量消化率都会影响色氨酸需要量[16]。表 10-1 中展示了 17 种水产动物的色氨酸需要量，不同养殖品种色氨酸需要量不同，研究结果的总体范围为 0.13%～0.48%。由表 10-1 可知，鲤色氨酸需要量为 0.13%，而麦瑞加拉鲮色氨酸需要量为 0.48%；即使是同一品种，不同规格、不同统计模型、不同参考指标，色氨酸的需要量也不同，如露斯塔野鲮、罗非鱼色氨酸需要量分别为 0.21%～0.38%、0.25%～0.38%。色氨酸缺乏或过量均抑制水产动物生长，降低消化酶活性及免疫功能[17,18]。少带重牙鲷分别投喂对照组饲料以及分别添加 2.1%赖氨酸或 0.4%色氨酸时，添加色氨酸组增重率显著低于对照组、赖氨酸组，其可能是由饲料总色氨酸过量造成的[19]。适量色氨酸可以提高水产动物肠道长度、重量、指数及褶皱高度，提高肝胰腺和肠道蛋白酶、脂肪酶及淀粉酶的活性，进而提高蛋白质、氨基酸、干物质的表观消化率[20,21]，同时提高其他必需氨基酸、非必需氨基酸的沉积率，进而促进水产动物生长[22]。胰岛素样生长因子-Ⅰ（IGF-Ⅰ）在水产动物生长、发育过程中发挥重要作用，色氨酸参与 IGF-Ⅰ的合成调控，适量色氨酸含量促进 IGF-Ⅰ的合成、分泌[23]。此外，成肌调节因子（MRFs）和肌生成抑制蛋白是水产动物肌肉生长的重要调节因子，这些调节因子受营养素的调控，特别是色氨酸具有促进肌肉生长的作用[24]。罗非鱼饲料中色氨酸含量为 0.37%时，具有活化增殖肌卫星细胞作用的肌分化因子和肌细胞生成素的基因表达量显著升高，具有抑制肌肉生长作用的肌生成抑制蛋白的基因表达量显著减低[25,26]。上述研究结果表明，适量色氨酸通过改善肠道形态结构、提高消化酶活性、促进 IGF-Ⅰ的合成及肌肉生长，从而提高水产动物的生长性能。

表 10-1 水产动物色氨酸需要量

品种	体重（g）	模型	指标	需要量（占饲料干物质，%）	参考文献
露斯塔野鲮	0.37	二次多项式模型	增重率	0.38	Abidi 等[6]
			蛋白质沉积率	0.36	
	1.50	折线模型	特定生长率	0.45	Murthy 等[27]
	3.45	非线性模型	绝对增重	0.25	Zehra 等[28]
			RNA/DNA	0.21	
罗非鱼	5.30	二次多项式模型	增重率	0.38	Prabu[26]
	3.40	二次多项式模型	增重率	0.34	Zaminhan 等[22]
			色氨酸保留率	0.34	
	7.40	折线模型	积温生长系数	0.31	Nguyen 等[29]
		非线性模型	饲料效率	0.25	
		非线性模型	蛋白质净保留率	0.27	
	38.20	二次多项式模型	增重率	0.29	Zaminhan 等[3]
			饲料效率	0.31	

(续)

品种	体重（g）	模型	指标	需要量（占饲料干物质，%）	参考文献
眼斑拟石首鱼	2.29	二次多项式模型	增重率	0.28	Pewitt 等[30]
			蛋白质效率	0.23	
团头鲂	23.33	折线模型	增重率	0.20	Ji 等[17]
杂交条纹鲈	42.40	非线性模型	增重率	0.21	Gaylord 等[5]
			IGF-I 含量	0.19	
印度囊鳃鲇	6.66	折线模型	增重率	0.28	And 等[31]
			蛋白质效率	0.29	
麦瑞加拉鲮	0.62	二次多项式模型	增重率	0.42	Ahmed 等[32]
			蛋白质效率	0.39	
	1.10	折线模型	增重率	0.48	Benakappa 等[33]
非洲鲇	11.5	非线性模型	增重率	0.44	Fagbenro 等[4]
虹鳟	14	折线模型	增重率	0.25	Walton 等[2]
建鲤	7.73	二次多项式模型	增重率	0.35	Tang 等[21]
草鱼	287	二次多项式模型	增重率	0.38	Wen 等[34]
		二次多项式模型	鳃 MDA 含量	0.41	Jiang 等[18]
印度鲇	4.44	二次多项式模型	增重率	0.37	Ahmed[35]
			蛋白质效率	0.32	
尖吻鲈	5.3	折线模型	增重率	0.21	Coloso 等[36]
遮目鱼	7.7	折线模型	增重率	0.31	Coloso 等[37]
大麻哈鱼	1.1		增重率	0.29	Akiyama 等[38]
鲤	0.20	非线性模型	增重率	0.13	Dabrowski[39]
凡纳滨对虾	0.31	折线模型	特定生长率	0.28	王用黎[40]

第四节 色氨酸对水产动物摄食量的影响

适量色氨酸能促进水产动物的摄食。罗非鱼高植物蛋白饲料中分别添加 0.4 g/kg 二甲基-β-丙酸噻亭、0.6 g/kg 二甲基乙酸噻亭（DMP）、1.8 g/kg 色氨酸、6 g/kg 甜菜碱，研究结果发现，色氨酸和 DMP 组的摄食量显著高于对照组，其主要原因是色氨酸促进脑神经肽 Y 的基因表达，DMP 促进胃饥饿素的基因表达[41]。印度囊鳃鲇饲料色氨酸含量 0.24% 和 0.27% 时，摄食量显著高于色氨酸含量 0.1% 和 0.15% 组[31]。牙鲆饲料中分别添加 0.12%、0.3% 和 0.5% 色氨酸时，摄食量随着色氨酸添加量的增加显著升高[42]。当色氨酸含量高于需要量时，随着色氨酸含量的增加，摄食量显著降低。建鲤饲料中色氨酸含量为 0.11%~0.38% 时，随着饲料色氨酸含量的增加，摄食量显著提高；色氨酸含量为 0.38%~0.69% 时，随着饲料色氨酸含量的增加，摄食量显著降低[21]。适量色氨酸含量提高水产动物的生产性能，可能与提高其摄食量有关[31,35]。草鱼饲料中色氨酸含量为 0.07%～0.31%时，摄食量和增重率随着饲料色氨酸含量的升高而显著提高，且增重率与摄食量呈显著正相关[34]。然而，罗非鱼研究结果显示，饲料色氨酸含量对摄食量没有显著影响[3]。色氨酸对摄食量影响不一致的可能原因是试验饲料中色氨酸的含量、试验饲料原料及品种差异等。

第五节 色氨酸对水产动物应激反应的影响

适量色氨酸能缓解亚硝酸盐、高铜、拉网、细菌感染、拥挤等造成的应激胁迫。露斯塔野鲮饲料中添加 0.75%或 1%色氨酸显著改善了亚硝酸盐应激导致的增重率、特定生长率显著降低以及饲料系数显著升高，肝脏、肌肉谷草转氨酶和谷丙转氨酶活性显著降低的问题，缓解了亚硝酸盐的应激反应[43]；此外，色氨酸可以缓解温度、盐度对露斯塔野鲮的应激反应[44]。拟鲤饲料中添加 0.25%色氨酸显著降低高铜应激的死亡率[45]。加利福尼亚湾石首鱼饲料添加需要量的 1 倍或 2 倍色氨酸时，拉网应激后端脑 5-羟基吲哚乙酸和 5-羟色胺含量显著升高；然而，添加需要量的 3 倍或 4 倍色氨酸时两者显著降低[46]。锯缘青蟹饲料中添加 0.75%或 1%色氨酸显著降低其攻击强度和频率，添加 0.5%、0.75%或 1%色氨酸显著提高血淋巴 5-羟色胺的含量和青蟹成活率，但增重率、特定生长率均显著降低[47]。

此外，为提高水产动物单位面积产量，高密度养殖是最常用的养殖模式，但高密度养殖通常会带来拥挤胁迫应激，导致皮质醇释放增加，长期胁迫影响水产动物健康[48,49]。高密度养殖条件下，海参特定生长率显著低于低密度养殖组，饲料中添加 1%或 3%色氨酸通过改变机体能量分配，提高用于生长的能量，降低机体皮质醇、葡萄糖及乳酸的含量，显著提高海参特定生长率[50]。麦瑞加拉鲮饲料中添加 2.72%色氨酸能够缓解高密度养殖对皮质醇含量及生长的负面影响[50]。虹鳟饲料中添加 0.5%色氨酸显著提高低密度养殖条件下的生长性能、肌肉粗脂肪含量，显著降低血清皮质醇、葡萄糖的含量；添加 0.5%或 1%色氨酸显著降低高密度养殖条件下虹鳟的饲料系数[52]。上述研究结果表明，色氨酸可以缓解高密度养殖带来的拥挤胁迫，提高水产动物生长性能。

色氨酸缓解应激胁迫的可能机制是：①色氨酸可以促进机体 5-羟色胺的生物合成，合成的 5-羟色胺进而直接抑制皮质醇的合成和攻击行为，进而缓解应激反应[8,15,53]。欧洲鲈美人鱼发光杆菌攻毒 48 h 和 72 h 后，色氨酸添加组血清中皮质醇含量显著低于色氨酸缺乏组，从而缓解攻毒应激反应[54]。②应激反应产生的皮质醇激活糖酵解和糖异生，产生大量葡萄糖用于抗应激，饲料中添加色氨酸可以缓解应激反应，降低血清葡萄糖的含量[55,56]。罗非鱼饲料中色氨酸含量为 0.37%和 0.48%时，血清中皮质醇和葡萄糖的含量均显著低于 0.26%组[26]。团头鲂饲料中不同含量色氨酸对血清葡萄糖含量的影响未达到显著水平，但随着添加量的增加血糖含量先降低后升高，色氨酸调控糖代谢的主要途径是降低糖酵解葡萄糖激酶的基因表达量和提高糖异生磷酸烯醇丙酮酸羧激酶、葡萄糖-6-磷酸酶相关基因的表达量[17]。

第六节 色氨酸对水产动物抗病力、抗氧化和免疫功能的影响

色氨酸代谢产物具有广泛的生理作用，除上述缓解应激外，色氨酸还有提高抗病力、抗氧化及机体免疫等作用。欧洲鲈摄食色氨酸含量缺乏饲料、色氨酸含量正常饲料、正常饲料添加0.13%和0.17%色氨酸饲料，经美人鱼发光杆菌攻毒8 d后，色氨酸含量缺乏组死亡率最高，其次是正常饲料添加0.17%色氨酸组，正常组和正常饲料添加0.13%色氨酸组死亡率最低，结果表明色氨酸缺乏或过量均降低欧洲鲈抗病力[54]。虹鳟在低密度和高密度养殖条件下，通过饲料中添加0.5%色氨酸显著提高血清溶菌酶活性、杀菌活性及过氧化氢酶（CAT）活性，显著降低血清丙二醛（MDA）的含量，提高机体抗氧化性能；高密度养殖条件下，添加0.5%色氨酸显著提高肠道免疫因子肿瘤坏死因子-α（$TNF-\alpha$）、抗炎因子白介素1-β（$IL1-\beta$）、白介素8（$IL8$）的基因表达量，提高抗氧化酶超氧化物歧化酶（SOD）、CAT、谷胱甘肽过氧化物酶（GPx）基因的表达量及活性；添加1%色氨酸提高$TNF-\alpha$的基因表达量及SOD活性，降低促炎因子白介素6（$IL6$）、SOD基因的表达量[57,58]。欧洲鲈饲料中添加0.5%色氨酸投喂15 d后，血细胞比容、血红蛋白、红细胞及白细胞数量与对照组相比均没有显著影响；美人鱼发光杆菌攻毒后，0.5%色氨酸组红细胞平均血红蛋白浓度、中性粒细胞值显著提高[59]。上述研究结果表明，色氨酸通过提高抗氧化酶活性及基因表达，抑制促炎因子、提高抗炎因子的基因表达，从而提高机体抗病力。

色氨酸调控水产动物抗病力、抗氧化、免疫功能的可能机制为：①色氨酸作为神经内分泌5-羟色胺和褪黑素的前体物，在水产动物应激条件下添加，能够增加二者的合成量，进而提高下游抗氧化酶的活性，起到治疗作用[61]。中华绒螯蟹饲料中添加色氨酸显著提高肝胰腺的SOD活性，这与褪黑素注射1 d后，肝胰腺的总抗氧化（T-AOC）、SOD活性显著提高的效果相似[61]。②色氨酸分解代谢IDO可诱导干扰素-γ、释放细胞因子、激活免疫细胞、提高机体抗氧化和抗菌活性，进而提高机体免疫功能[12]。③色氨酸通过核因子E2相关因子2/Kelch样环氧氯丙烷相关蛋白α（Nrf2/Keapα）信号通路提高抗氧化酶活性及其基因表达量，通过磷脂酰肌醇-3-羟激酶-雷帕霉素靶蛋白（PI3K-TOR）、核转录因子κB（NF-κB）信号通路降低促炎因子的基因表达量，提高抑炎因子的基因表达量[62]。适量色氨酸水平提高草鱼抗炎因子白介素-10（$IL10$）、转化生长因子-β（$TGF-\beta$）、核转录因子抑制因子κBα（iκBα）的基因表达量，降低促炎因子$TNF-\alpha$、$IL8$及$IL8\beta$的基因表达量；提高SOD、CAT、GPx、谷胱甘肽还原酶（GR）、谷胱甘肽-S-转移酶（GST）的基因表达量[18]。④色氨酸调控机体健康还与肠道免疫有关，中华绒螯蟹饲料中添加0.47%或0.73%色氨酸，显著提高嗜水气单胞菌攻毒后成活率，提高肠道菌群丰度和多样性指数，变形菌门、厚壁菌门、放线菌门优势菌群丰度的提高，改善了肠道健康[63]。草鱼饲料添加色氨酸显著提高铜胁迫后肠道溶菌酶、酸性磷酸酶活性及补体C3的含量，提高抗氧化酶SOD、GPX的活性及谷胱甘肽的含量，降低MDA、蛋白质羰基含量；通过提高肠道$TGF-\beta$、$IL10$、闭合蛋白、闭锁小带、SOD及GPx的基因表达量，降低$IL8$、$TNF-\alpha$、$Keap1$等基因表达量，改善肠道健康，提高机体抗病力[34]。

第七节 小结与展望

目前，关于水产动物色氨酸的需要量及营养生理功能已有大量研究，现有水产动物色氨酸的需要量为 0.13%～0.48%（占饲料百分比），适量的色氨酸水平可以提高水产动物摄食量，缓解水产动物对高养殖密度、高铜、亚硝酸盐等的应激反应，通过提高免疫、抗氧化功能，进而提高抗病力。但是，关于色氨酸在水产动物体内的代谢途径、色氨酸对机体糖和脂肪等代谢的影响机制、大规格水产动物对色氨酸的需要量等方面仍需开展更为全面的研究。

参考文献

[1] Lall S P, Davis D A. Technology and Nutrition [C] //Davis D A. Feed and feeding practices in Aquaculture. Cambridge: Woodhead Publishing, 2015: 287.

[2] Walton M J, Coloso R M, Cowey C B, et al. The effects of dietary tryptophan levels on growth and metabolism of rainbow trout (*Salmo gairdneri*) [J]. British Journal of Nutrition, 1984, 51: 279-287.

[3] Zaminhan M, Boscolo W, Neu D H, et al. Dietary tryptophan requirements of juvenile Nile tilapia fed corn-soybean meal-based diets [J]. Animal Feed Science and Technology, 2017, 227: 62-67.

[4] Fagbenro O A, Nw Anna L C. Dietary tryptophan requirement of the African catfish, *Clarias gariepinus* [J]. Journal of Applied Aquaculture, 1999, 9 (1): 65-72.

[5] Gaylord T G, Rawles S D, Davis K B. Dietary tryptophan requirement of hybrid striped bass (*Morone chrysops* × *M. saxatilis*) [J]. Aquaculture Nutrition, 2005, 11: 367-374.

[6] Abidi S F, Khan M A. Dietary Tryptophan Requirement of Fingerling Rohu, *Labeo rohita* (Hamilton), Based on Growth and Body Composition [J]. Journal of the World Aquaculture Society, 2010, 41 (5): 700-709.

[7] Sun Y P, Guan L C, Xiong J H, et al. Effects of L-tryptophan-supplemented dietary on growth performance and 5-HT and GABA levels in juvenile *Litopenaeus vannamei* [J]. Aquaculture International, 2014, 23 (1): 235-251.

[8] Martins C I M, Silva P I M, Costas B, et al. The effect of tryptophan supplemented diets on brain serotonergic activity and plasma cortisol under undisturbed and stressed conditions in grouped-housed Nile tilapia *Oreochromis niloticus* [J]. Aquaculture, 2013, 400-401: 129-134.

[9] Basic D, Krogdahl A, Schjolden J, et al. Short- and long-term effects of dietary L-tryptophan supplementation on the neuroendocrine stress response in seawater-reared Atlantic salmon (*Salmo salar*) [J]. Aquaculture, 2013, 388-391: 8-13.

[10] Hseu J R, Lu F I, Su H M, et al. Effect of exogenous tryptophan on cannibalism, survival and growth in juvenile grouper, *Epinephalus coioides* [J]. Aquaculture, 2003, 218: 251-263.

[11] Harlioğlu M M, Harlioğlu A G, Mişe Y S, et al. Effects of dietary L-tryptophan on the agonistic behavior, growth, and survival of freshwater crayfish *Astacus leptodactylus* Eschscholtz [J]. Aquaculture International, 2014, 22: 733-748.

[12] Le Floc'h N, Otten W, Merlot E, et al. Tryptophan metabolism, from nutrition to potential

therapeutic applications [J]. Amino Acids, 2011, 41: 1195-1205.

[13] Cortes J, Alvarez C, Santana P, et al. Indoleamine 2, 3 - dioxygenase: First evidence of expression in rainbow trout (*Oncorhynchus mykiss*) [J]. Developmental and Comparative Immunology, 2016, 65: 73-78.

[14] Ng W K, Serrini G, Zhang Z, et al. Niacin requirement and inability of tryptophan to act as a precursor of NAD^+ in channel catfish, *Ictalurus punctatus* [J]. Aquaculture, 1997, 152: 273-285.

[15] Hoseini S M, Pérez-Jiménez A, Costas B, et al. Physiological roles of tryptophan in teleosts: current knowledge and perspectives for future studies [J]. Reviews in Aquaculture, 2019, 11: 3-24.

[16] De Silva S S, Gunasekera R M, Gooley G. Digestibility and amino acid availability of three protein - rich ingredient incorporated diets by Murray cod *Maccullochell apeelii* (Mitchell) and the Australian short fin eel *Anguilla australis* Richardson [J]. Aquaculture Research, 2000, 31: 195-205.

[17] Ji K, Liang H, Chisomo - Kasiya H, et al. Effects of dietary tryptophan levels on growth performance, whole body composition and gene expression levels related to glycometabolism for juvenile blunt snout bream, *Megalobrama amblycephala* [J]. Aquaculture Nutrition, 2018, 24: 1474-1483.

[18] Jiang W D, Wen H L, Liu Y, et al. The tight junction protein transcript abundance changes and oxidative damage by tryptophan deficiency or excess are related to the modulation of the signalling molecules, NF - kappa B p65, TOR, caspase - (3, 8, 9) and Nrf2 mRNA levels, in the gill of young grass carp (*Ctenopharyngodon idellus*) [J]. Fish & Shellfish Immunology, 2015, 46 (2): 168-180.

[19] Saavedra M, Barr Y, Pedro P F, et al. Supplementation of tryptophan and lysine in *Diplodus sargus* larval diet: effects on growth and skeletal deformities [J]. Aquaculture Research, 2010, 40 (10): 1191-1201.

[20] 孙育平, 裘金木, 王国霞, 等. 低蛋白质饲料中添加色氨酸对凡纳滨对虾饲料表观消化率、消化酶活和全虾氨基酸组成的影响 [J]. 水生生物学报, 2016, 40 (4): 720-727.

[21] Tang L, Feng L, Sun C Y, et al. Effect of tryptophan on growth, intestinal enzyme activities and TOR gene expression in juvenile Jian carp (*Cyprinus carpio* var. Jian): Studies in vivo and in vitro [J]. Aquaculture, 2013, 412: 23-33.

[22] Zaminhan M, Michelato M, Furuya Vrb, et al. Total and available tryptophan requirement of Nile tilapia, *Oreochromis niloticus*, fingerlings [J]. Aquaculture Nutrition, 2018, 24: 1553-1562.

[23] Dyer A R, Barlow C G, Bransden M P, et al. Correlation of plasma IGF - I concentration and growth rate in aquacultured finfish: a tool for assessing the potential of new diets [J]. Aquaculture, 2004, 236: 583-592.

[24] Johnston I A, Lee H T, Macqueen D J, et al. Embryonic temperature affects muscle fibre recruitment in adult zebrafish: Genome - wide changes in gene and microRNA expression associated with the transition from hyperplastic to hypertrophic growth phenotypes [J]. Journal of Experimental Biology, 2009, 212: 1781-1793.

[25] Lee C Y, Hu S Y, Gong H Y, et al. Suppression of myostatin with vector - based RNA interference causes a double - muscle effect in transgenic zebrafish [J]. Biochemical and Biophysical Research Communications, 2009, 387: 766-771.

[26] Prabu E, Felix N, Uma A, et al. Metabolic responses of juvenile GIFT strain of Nile tilapia (*Oreochromis niloticus*) to dietary L - tryptophan supplementation [J]. Aquaculture Nutrition, 2020, 26: 1713-1723.

[27] Murthy H S, Varghese T J Dietary Tryptophan Requirement for the Growth of Rohu, *Labeo rohita* [J]. Journal of Applied Aquaculture, 1997, 7 (2): 71-79.

[28] Zehra S, Khan M A. Dietary tryptophan requirement of fingerling *Catla catla* (Hamilton) based on growth, protein gain, RNA/DNA ratio, haematological parameters and carcass composition [J]. Aquaculture Nutrition, 2015, 21: 690-701.

[29] Nguyen L, Salem S M R, Salze G P, et al. Tryptophan requirement in semi-purified diets of juvenile Nile tilapia *Oreochromis niloticus* [J]. Aquaculture, 2018, 502: 258-267.

[30] Pewitt E, Castillo S, Alejandro Velasqez, et al. The dietary tryptophan requirement of juvenile red drum, *Sciaenops ocellatus* [J]. Aquaculture, 2017, 469: 112-116.

[31] And F, Khan M A. Dietary L-tryptophan requirement of fingerling stinging catfish, *Heteropneustes fossilis* (Bloch) [J]. Aquaculture Research, 2014, 45: 1224-1235.

[32] Ahmed I, Khan M A. Dietary tryptophan requirement of fingerling Indian major carp, *Cirrhinus mrigala* (Hamilton) [J]. Aquaculture Research, 2005, 36 (7): 687-695.

[33] Benakappa S, Varghese T. Dietary requirement of tryptophan for growth and survival of the Indian major carp, *Cirrhinus mrigala* (Hamilton-Buchanan) fry [J]. Indian Journal of Experimental Biology, 2003, 41: 1342-1345.

[34] Wen H, Feng L, Jiang W, et al. Dietary tryptophan modulates intestinal immune response, barrier function, antioxidant status and gene expression of TOR and Nrf2 in young grass carp (*Ctenopharyngodon idella*) [J]. Fish & Shellfish Immunology, 2014, 40 (1): 275-287.

[35] Ahmed. Dietary amino acid l-tryptophan requirement of fingerling Indian catfish, *Heteropneustes fossilis* (Bloch), estimated by growth and haemato-biochemical parameters [J]. Fish Physiology and Biochemistry, 2012, 38: 1195-1209.

[36] Coloso R, Murillo-Gurrea D, Borlongan I, et al. Tryptophan requirement of juvenile Asian sea bass *Lates calcarifer* [J]. Journal of Applied Ichthyology, 2004, 20: 43-47.

[37] Coloso R M, Tiro L B, Benttez L V, et al. Requirement for tryptophan by milkfish (*Chanos chanos* Forsskal) juveniles [J]. Fish Physiology and Biochemistry, 1992, 10: 35-41.

[38] Akiyama T, Arai S, Murai T, et al. Tryptophan requirement of chum salmon fry [J]. Nippon Suisan Gakkaishi, 1985, 51: 1005-1008.

[39] Dabrowski. Tryptophan requirement of common carp (*Cyprinus carpio* L.) fry [J]. Zeitschriftfür Tierphysiologie Tierernährung und Futtermittelkunde, 1981, 46: 64-71.

[40] 王用黎. 凡纳滨对虾幼虾对苏氨酸、亮氨酸、色氨酸和缬氨酸需要量的研究 [D]. 湛江: 广东海洋大, 2013: 31-32.

[41] Zou Q, Huang Y, Cao J, et al. Effects of four feeding stimulants in high plant-based diets on feed intake, growth performance, serum biochemical parameters, digestive enzyme activities and appetite-related genes expression of juvenile GIFT tilapia (*Oreochromis sp.*) [J]. Aquaculture Nutrition, 2017, 23: 1076-1085.

[42] Han Y H, Koshio S, Ishikawa M, et al. Optimum Supplementations and Interactive Effects of Methionine and Tryptophan on Growth and Health Status of Juvenile Japanese Flounder *Paralichthys olivaceus* [J]. Aquaculture Science, 2013, 61 (3): 239-251.

[43] Cui A, Sahu N P, Pal A K, et al. Dietary L-tryptophan modulates growth and immuno-metabolic status of *Labeo rohita* juveniles exposed to nitrite [J]. Aquaculture Research, 2015, 46 (8): 2013-2024.

[44] Akhtar M, Pal A, Sahu N, et al. Physiological responses of dietary tryptophan fed *Labeo rohita* to temperature and salinity stress [J]. Journal of animal physiology and animal nutrition, 2013, 97: 1075-1083.

[45] Fattahi S, Hoseini S M. Effect of dietary tryptophan and betaine on tolerance of Caspian roach (*Rutilus rutilus caspicus*) to copper toxicity [J]. International Journal of Aquatic Biology, 2013, 1 (2): 76-81.

[46] Miguel C G, Lopez Lus M, Galaviz M A, et al. Effect of L-tryptophan supplemented diets on serotonergic system and plasma cortisol in *Totoaba macdonaldi* (Gilbert, 1890) juvenile exposed to acute stress by handling and hypoxia [J]. Aquaculture Research, 2018, 49: 847-857.

[47] Jr J L Q L, Quinttio E T, Catacutan M R, et al. Effects of dietary L-tryptophan on the agonistic behavior, growth and survival of juvenile mud crab *Scylla serrata* [J]. Aquaculture, 2010, 310 (1): 84-90.

[48] Yousefi M, Paktinat M, Mahmoudi N, et al. Serum biochemical and non-specific immune responses of rainbow trout (*Oncorhynchus mykiss*) to dietary nucleotide and chronic stress [J]. Fish Physiology and Biochemistry, 2016, 42: 1417-1425.

[49] Taheri M A, Hoseini S M, Ghelichpour M, et al. Effects of dietary 1, 8-cineole supplementation on physiological, immunological and antioxidant responses to crowding stress in rainbow trout (*Oncorhynchus mykiss*) [J]. Fish and Shellfish Immunology, 2018, 81: 182-188.

[50] Zhang E, Dong S, Wang F, et al. Effects of L-tryptophan on the performance, energy partitioning and endocrine response of Japanese sea cucumber (*Apostichopus japonicus* Selenka) exposed to crowding stress [J]. Aquaculture Research, 2018, 49: 471-479.

[51] Tejpal C S, Pal A K, Sahu N P, et al. Dietary supplementation of L-tryptophan mitigates crowding stress and augments the growth in *Cirrhinus mrigala* fingerlings [J]. Aquaculture, 2009, 293: 272-277.

[52] Hoseini S M, Mirghaed A T, Ghelichpour M, et al. Effects of dietary tryptophan supplementation and stocking density on growth performance and stress responses in rainbow trout (*Oncorhynchus mykiss*) [J]. Aquaculture, 2020, 519: 734908.

[53] Hoseini S M, Hosseini S A. Effect of dietary L-tryptophan on osmotic stress tolerance in common carp, *Cyprinus carpio*, juveniles [J]. Fish Physiology and Biochemistry, 2010, 36: 1061-1067.

[54] Machado M, Azeredo R, Domingues A, et al. Dietary tryptophan deficiency and its supplementation compromises inflammatory mechanisms and disease resistance in a teleost fish [J]. Scientific Reports, 2019, 9: 7689.

[55] Hoseini S M, Hosseini S A, Soudagar M. Dietary tryptophan changes serum stress markers, enzyme activity, and ions concentration of wild common carp *Cyprinus carpio* exposed to ambient copper [J]. Fish Physiology and Biochemistry, 2012, 38: 1419-1426.

[56] Lepage O, Tottmar O, Winberg S. Elevated dietary intake of L-tryptophan counteracts the stress-induced elevation of plasma cortisol in rainbow trout (*Oncorhynchus mykiss*) [J]. Journal of Experimental Biology, 2002, 205: 3679-3687.

[57] Hoseini S M, Taheri M A, Ghelichpour M. Effects of dietary tryptophan levels and fish stocking density on immunological and antioxidant responses and bactericidal activity against *Aeromonas hydrophila* in rainbow trout (*Oncorhynchus mykiss*) [J]. Aquaculture Research, 2020, 51: 1455-1463.

[58] Hoseini S M, Yousefi M, Mirghaed A T, et al. Effects of rearing density and dietary tryptophan

supplementation on intestinal immune and antioxidant responses in rainbow trout (*Oncorhynchus mykiss*) [J]. Aquaculture, 2020, 528: 735537.

[59] Machado M, Azeredo R, Díaz-Rosales P, et al. Dietary tryptophan and methionine as modulators of European seabass (*Dicentrarchus labrax*) immune status and inflammatory response [J]. Fish and Shellfish Immunology, 2015, 42: 353-362.

[60] Esteban M A, Rodriguez A, Ayala A G, et al. Effects of high doses of cortisol on innate cellular immune response of seabream (*Sparus aurata* L.) [J]. General and Comparative Endocrinology, 2004, 137: 89-98.

[61] 徐敏杰，张佳鑫，黄根勇，等. L-色氨酸和褪黑激素对中华绒螯蟹血清血糖水平及肝胰腺抗氧化能力的影响 [J]. 水产学报, 2018, 42 (01): 91-99.

[62] Ji K, Liang H, Ren M, et al. Effects of dietary tryptophan levels on antioxidant status and immunity for juvenile blunt snout bream (*Megalobrama amblycephala*) involved in Nrf2 and TOR signaling pathway [J]. Fish & Shellfish Immunology, 2019, 93: 474-483.

[63] Yang X, Xu M, Huang G, et al. Effect of dietary L-tryptophan on the survival, immune response and gut microbiota of the Chinese mitten crab, *Eriocheir sinensis* [J]. Fish & Shellfish Immunology, 2019, 84: 1007-1017.

第十一章 苯丙氨酸

苯丙氨酸（phenylalanine，Phe）是鱼类不能靠自身代谢合成的几种必需氨基酸之一，属芳香族氨基酸。在人体内，Phe能够直接合成重要的神经递质和激素，参与人体供能所必需的糖脂代谢活动[1,2]；在植物体内，Phe可对细胞器和组织发育起到直接作用，促进植物细胞的生长与信号转导[3]；在医药领域，苯丙氨酸还可作为抗癌药物的载体将药物分子直接导入癌瘤区，有效抑制癌细胞的分化，减少药物的毒副作用[4]；在食品加工领域，苯丙氨酸能够和糖类等物质发生一系列酰胺化反应，改善食物风味，维持人体日常营养所需的氨基酸平衡[5,6,7]。然而，苯丙氨酸在水产动物中的生理活性机制尚不明晰。本章综述了苯丙氨酸的合成与代谢、需要量及其在水产动物中的应用，以期为苯丙氨酸在水产动物的营养需要、代谢机制与绿色养殖等方面的进一步研究提供重要基础。

第一节 苯丙氨酸的理化性质

苯丙氨酸，化学名2-氨基苯丙酸，属芳香族氨基酸，分子式$C_9H_{11}NO_2$，分子质量165.19，熔点为283 ℃，分子结构如图11-1所示，常温下为白色结晶或结晶性粉末固体，减压升华，溶于水，难溶于甲醇、乙醇、乙醚。

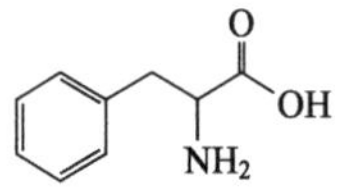

图11-1 苯丙氨酸分子结构

第二节 苯丙氨酸的合成与代谢途径

一、苯丙氨酸的生物合成

苯丙氨酸的合成起始物质是赤藓糖-4-磷酸（E4P）和磷酸烯醇式丙酮酸（PEP），二者经过一系列氧化、脱氢、脱羧、分子重排与缩合形成3-脱氧-α-阿拉伯庚酮糖酸-7-磷酸（DAHP）。然后，DAHP在脱氢奎宁酸（DHQ）合成酶、DHQ脱水酶和莽草酸脱氢酶的作用下，生成莽草酸。莽草酸是苯丙氨酸合成的前体物质。以莽草酸为起始物，经多种酶作用并最终形成分支酸的过程，称为莽草酸途径[8]。这一途径是芳香族氨基酸合成的共同步骤。随后，作为芳香族氨基酸合成途径的分支点，在分支酸变位酶与预苯酸脱水酶的作用下，先后转变成预苯酸和苯丙酮酸，最后在转氨酶的作用下形成苯丙氨酸[9]（图11-2）。

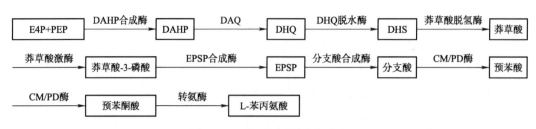

图 11-2 苯丙氨酸的生物合成

注：E4P 表示赤藓糖-4-磷酸，PEP 表示磷酸烯醇式丙酮酸，DAHP 表示 3-脱氧-a-阿拉伯庚酮糖酸-7-磷酸，DAQ 表示脱氢奎宁酸合成酶，DHQ 表示脱氢奎宁酸，DHS 表示脱氢莽草酸，EPSP 表示 5-烯醇丙酮酸莽草酸-3-磷酸，CM/PD 酶表示分支酸变位酶/P-预苯酸脱水酶。

二、苯丙氨酸的代谢

苯丙氨酸在生物体内的主要代谢途径是通过羟化反应生成酪氨酸（Tyrosine，Tyr）[10]（图 11-3）。苯丙氨酸可在辅酶四氢生物蝶呤的作用下不可逆地转化为 Tyr[11]。当体内苯丙氨酸羟化酶发生缺陷时，苯丙氨酸不能正常转变为 Tyr，苯丙氨酸在转氨基作用下生成苯丙酮酸、苯乙酸等，且随尿排出，俗称苯酮酸尿症（Phenylketonuria，PKU），是最常见的氨基酸代谢缺陷[12,13]。苯丙氨酸缺乏时，会影响动植物体内 Tyr 合成，并造成甲状腺素水平下降，影响动植物的正常代谢活动，严重时可导致智力低下、癫痫等[14]。

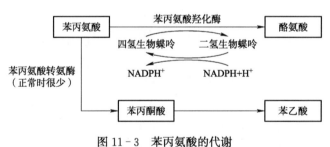

图 11-3 苯丙氨酸的代谢

第三节 水产动物苯丙氨酸需要量

不同种类水产动物对苯丙氨酸的需要量见表 11-1。由表可知，水产动物对苯丙氨酸需要量的范围变化较大，水产动物对苯丙氨酸的需要量与其种类、大小、饲料类型、饲养管理条件、投喂方法、饲料氨基酸组成、评价指标和统计方法等密切相关。同一品种不同饲养规格的水产动物对饲料中苯丙氨酸的需要量不同。当草鱼体重为 255.97 g 时，饲料干粮中苯丙氨酸需要量为 1.04%[15]，而体重为 13.21 g 的草鱼饲料干粮苯丙氨酸需要量在 1.22%～1.27%[16]。当体重为 1.65 g 和 158.78 g 时，尼罗罗非鱼对苯丙氨酸的需要量分别为 1.21% 和 3.66%[17,18]。在露斯塔野鲮研究中，露斯塔野鲮鱼苗苯丙氨酸需要量为 1.48%，而体重为 0.18 g 和 0.92 g 的露斯塔野鲮对苯丙氨酸需要量则分别为 1.16% 和 1.75%[19,20]。当体重为 0.58 g 时，遮目鱼对苯丙氨酸需要量为 1.26%～1.90%；当体重达到 8.00 g 时，其苯丙氨酸需要量为 1.12%～1.69%[21,22]。在日本鳗鲡对饲料苯丙氨酸需要量的研究中，黑子鳗为 1.89%，幼鳗为 1.76%，成鳗为 1.60%，以上需要量均低于 NRC 中日本鳗鲡成鳗在正常饲养水平下的必需氨基酸推荐值[23]。不同评价指标对水产动物苯丙氨酸需要量也存在一定差异。在草鱼的研究中表明，以增重率、饲料效率和蛋白质沉积率为评价指标，通过折线回归分析模型得出，体重为 13.21g 的草鱼对苯丙氨酸的需要量为 1.22%～1.27%[16]。Kim 等[24]以真鲷为研究对象，评价了苯丙氨酸二肽形式作为一种新型氨基酸来源的效果。研究结果表明，二肽苯丙氨酸对真鲷幼鱼的利用与自由形态一样有效，对其生长性能和全身氨基酸组成均无不良影响。Baker[25]在对不同统计方法进行比较分析后得出，通常条件下由折线法计算得出的氨基酸需要量往往比非线性方法估计得出的氨基酸需要量低。

表 11-1　水产动物苯丙氨酸需要量

物种	体重（g）	模型	指标	需要量（占饲料干物质，%）			参考文献
				Phe	Tyr	Phe+Tyr	
斑点叉尾鮰	195.00～205.00	二次回归分析	增重率、饲料效率	0.47～0.49	0.60	—	Robinson 等[26]
草鱼	255.97	二次回归分析	增重率	1.04	1.07	—	Li 等[15]
草鱼	13.21	折线回归分析	增重率、饲料效率、蛋白质沉积率	1.22～1.27	0.55	—	孙丽慧等[16]
大鳞鲃	11.29	A/E 估计	鱼体氨基酸	0.85	—	—	许红等[27]
大鳞大麻哈鱼	—	二次回归分析	增重率	1.70	0.40	—	Chance 等[28]
大菱鲆	55.00	A/E 估计	鱼体氨基酸	—	—	5.30	Kaushik 等[29]
大西洋鲑	1.38	折线回归分析	IAA 效率	—	—	5.60	Rollin 等[30]
高首鲟	35.30	A/E 估计	鱼体氨基酸	—	—	6.57	Xu 等[31]
高首鲟	66.70	A/E 估计	鱼体氨基酸	—	—	15.50	Ng 等[32]
虹鳟	13.00	折线模型	增重率	0.70	1.33	—	Kim 等[33]
虹鳟	13.00	线性分析	—	—	—	1.50	Kim 等[33]
吉富罗非鱼	52.70	二次回归分析	增重率、特定生长率、蛋白质效率、蛋白质沉积率、饲料系数	1.17～1.21	—	—	蒋明等[34]
大西洋鲷	40.00	A/E 估计	鱼体氨基酸	—	—	2.90	Kaushik 等[29]
美洲鳗鲡	283.00～286.00	A/E 估计	肌肉氨基酸	1.9～2.2	—	—	何英霞等[35]
日本鳗鲡	52.00～520.00	A/E 估计	肌肉氨基酸	1.89～2.2	—	—	许琪娅等[23]
尼罗罗非鱼	—	折线回归法	—	1.55	0.50	—	Santiago 等[36]
尼罗罗非鱼	150～200	缺失法	鱼体氨基酸	3.66	—	—	Do Nascimento 等[17]
尼罗罗非鱼	1.65	二次回归分析	绝对增重、蛋白质沉积率	1.21	1.00	—	Zehra 等[18]
欧洲黑鲈	3.30	A/E 估计	鱼体氨基酸	—	—	2.60	Kaushik 等[29]

(续)

物种	体重（g）	模型	指标	需要量（占饲料干物质,%）			参考文献
				Phe	Tyr	Phe+Tyr	
牙鲆	3.00	A/E估计	鱼体氨基酸	—	—	3.80	Forster等[37]
日本囊对虾	0.79	A/E估计	体成分	1.30	—	—	Teshima等[38]
露斯塔野鲮	0.18	二次回归分析	特定生长率、饲料效率、蛋白质效率、体蛋白沉积	1.16~1.22	1.00	—	Khan等[19]
团头鲂	10.20	二次回归分析	特定生长率、饲料效率	0.93~1.01	1.07	≤3.44	Ren等[39]
乌鳢	5.96	A/E估计	肌肉氨基酸	—	—	6.29	尹东鹏等[40]
异育银鲫	3.19	非线性回归分析	特定生长率	1.09	—	—	马志英等[41]
银鲈	10.00	折线回归法	增重率	2.28	—	—	Ngamsnae等[42]
麦瑞加拉鲮	0.92	断点回归分析	增重率	1.75	0.52	2.27	Benakappa等[20]
卡特拉鲃	鱼苗	折线回归法	增重率	1.48	0.40		Ravi等[43]
遮目鱼	8.00	断点线性回归分析	饲料效率	1.69	0.40		Borlongan等[21]
遮目鱼	8.00	断点线性回归分析	饲料效率	1.12	1.07		Borlongan等[21]
遮目鱼	0.58	断点线性回归分析	增重率	1.90	0.45		Borlongan等[22]
遮目鱼	0.58	断点线性回归分析	增重率	1.26	1.20		Borlongan等[22]
真鲷	1.70	A/E估计	鱼体氨基酸	—	—	4.10	Forster等[37]

根据Kaushik等[44]研究方法，将增重率进行标准化相应处理后，对水产动物苯丙氨酸需要量与增重率进行Meta分析，得到二次多项式回归分析模型，即$y_{饲料干物质}=-36.434x^2+123.36x-0.2752$，$R^2=0.9813$；$y_{饲料蛋白}=-3.8923x^2+41.083x-4.16$，$R^2=0.9521$。以增重率为评价指标，经计算得到水产动物对苯丙氨酸的需要量占饲料干物质的1.69%，占饲料蛋白质的5.28%（图11-4）。

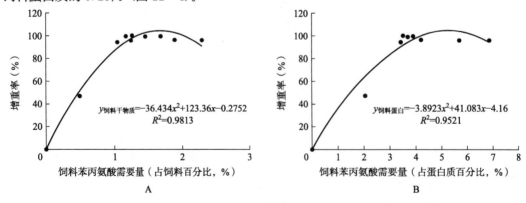

图11-4 水产动物饲料（干物质）苯丙氨酸与增重率的Meta分析

第四节 苯丙氨酸与其他营养素的相互作用

苯丙氨酸是酪氨酸的前体，酪氨酸参与多种分子的合成，且在代谢、生长、应激反应和色素沉着等方面发挥关键作用。由于苯丙氨酸可以转化为酪氨酸，饲料中酪氨酸水平对苯丙氨酸的需要量影响极大。研究表明，当酪氨酸水平为1.33%时，虹鳟幼鱼对苯丙氨酸的需要量为0.7%；而当饲料中同时添加酪氨酸和苯丙氨酸时，二者的需要量则为1.5%[33]。以增重率为评价指标，当酪氨酸水平为0.45%时，遮目鱼对苯丙氨酸的需要量为1.90%，而当饲料中酪氨酸水平为1.20%时，遮目鱼对苯丙氨酸的需要量则下降到1.26%[22]。Khan等[19]研究表明，当露斯塔野鲮幼鱼饲料酪氨酸水平为1.0%时，其对苯丙氨酸需要量为1.22%；而当饲料中缺乏酪氨酸时，露斯塔野鲮对苯丙氨酸的需要量为2.5%。此外，在日本鳗鲡的研究中发现，当饲料中缺乏酪氨酸时，日本鳗鲡对苯丙氨酸的需要量是2.2%；而当饲料中酪氨酸水平添加至2%时，日本鳗鲡对苯丙氨酸的需要量是1.2%[23]。由此可见，饲料中的酪氨酸对苯丙氨酸存在一定的节约作用。

此外，Saavedra等[45]研究表明，添加苯丙氨酸和酪氨酸的银鲷幼鱼对温度胁迫的存活率显著提高。同时，在饲料中补充苯丙氨酸和酪氨酸有助于减少应激引起的骨骼畸形和死亡率。在银鲷的研究中，氨基酸的需要量足以满足其生长需要，但不足以供应其他代谢过程。综上所述，饲料中酪氨酸和苯丙氨酸的补充对银鲷的生长和存活具有一定的积极作用。在温度胁迫条件下，苯丙氨酸降低了银鲷幼鱼脊椎椎体的压缩率，酪氨酸增加了银鲷幼鱼的抵抗力，这也意味着其他代谢过程对氨基酸的需要可能高于生长所需的氨基酸。

第五节 苯丙氨酸对水产动物蛋白质代谢及健康的影响

一、苯丙氨酸对水产动物蛋白质沉积率的影响

苯丙氨酸作为一种必需氨基酸，对水产动物的生长与代谢有着积极的促进作用。研究表明，饲料中适量的苯丙氨酸可显著提高虹鳟[33]和银鲈[42]等鱼类的生长性能。而苯丙氨酸的过量或缺乏都将影响水产动物对氨基酸的吸收利用，破坏饲料的氨基酸平衡，阻碍水产动物对蛋白质的合成，进而抑制其生长[41]。在对草鱼、麦瑞加拉鲮[46]、吉富罗非鱼[34]、团头鲂[39]、卡特拉鲃[43,47]和露斯塔野鲮[19]的研究中发现，饲料苯丙氨酸的过量或缺乏都将导致饲料系数的升高，引起水产动物生长性能的下降。Zehra等[18]在尼罗罗非鱼研究中发现，增重率和蛋白沉积率随饲料中苯丙氨酸含量的增加而增加，并在1.15%时出现最大值，进一步增加苯丙氨酸含量则导致尼罗罗非鱼增重率和蛋白沉积率显著下降。

值得注意的是，水产动物的种类和饲料酪氨酸水平等条件的不同都将导致不同结果。孙丽慧等[16]研究表明，当饲料干物质苯丙氨酸水平从0.82%提高到1.2%时，草鱼增重率、蛋白质沉积率和饲料效率显著升高，继续增加饲料苯丙氨酸水平后，草鱼生长性能则无显著变化。该结果与遮目鱼[22]和大鳞大麻哈鱼[28]等研究结果相似。然而，Ahmed等在研究麦瑞加拉鲮[48]时发现，在适当范围内降低饲料苯丙氨酸水平不会引起麦瑞加拉鲮生长性能的下降；但当饲料苯丙氨酸水平过量时，则会显著降低麦瑞加拉鲮增重率与蛋白沉

积率等生长性能。这可能意味着过量苯丙氨酸水平的抑制能力要强于缺乏水平。马志英等[41]研究表明,当饲料酪氨酸水平为1.04%且苯丙氨酸含量不足时,会引起异育银鲫幼鱼的生长性能下降和饲料系数的升高;当增加饲料苯丙氨酸水平后,其生长性能随着饲料中苯丙氨酸含量的增加而增加,超过一定范围后则导致异育银鲫生长性能显著下降。饲料中过高剂量的苯丙氨酸可能导致苯丙酮酸等代谢产物大量生成,并在水产动物体内聚积而产生毒性,对鱼类生长有一定的负面作用,抑制水产动物的生长[49,50],但具体机制需进一步研究证明。

二、苯丙氨酸对水产动物体成分的影响

国内外关于饲料苯丙氨酸水平对水产动物体成分的影响已有一些报道。Xiao等[51]研究发现,饲料苯丙氨酸水平为1.30%的杂交罗非鱼的全鱼蛋白质和脂肪含量显著高于苯丙氨酸水平为0.43%的鱼,而全鱼水分含量呈下降趋势。Kim等[33]研究表明,饲料苯丙氨酸水平可显著提高虹鳟鱼体粗蛋白质和粗脂肪含量。Khan等[19]对露斯塔野鲮全鱼体成分的研究表明,饲料中适宜水平的苯丙氨酸可显著增加露斯塔野鲮的粗脂肪和粗灰分含量。Ren等[39]在团头鲂饲料干粮中添加0.57%~2.04%水平的苯丙氨酸后发现,饲料苯丙氨酸水平可显著提高团头鲂全鱼水分、粗蛋白质和粗脂肪含量,但对全鱼粗灰分含量无显著影响。作为一种生酮氨基酸,苯丙氨酸可随饲料水平的增加而分解,为脂肪等生物大分子的合成提供足够的碳骨架,并导致全鱼粗脂肪含量的增加[52]。与露斯塔野鲮一样,吉富罗非鱼全鱼粗灰分含量随饲料苯丙氨酸水平的升高而升高,当苯丙氨酸水平达到1.72%后,全鱼粗灰分含量显著下降。而马志英等[41]研究则认为,饲料苯丙氨酸水平对异育银鲫的全鱼体成分无显著影响。孙丽慧等[16]发现,草鱼全鱼粗蛋白质含量随饲料苯丙氨酸水平显著升高,而全鱼水分、粗脂肪和粗灰分等含量则不随饲料苯丙氨酸水平的变化而变化。由以上结果可知,不同试验的研究结果不尽相同,除与水产动物的种类有关外,还与不同生长阶段、苯丙氨酸饲喂水平等因素有关。

三、苯丙氨酸对水产动物血清生化指标的影响

血清生化指标是一种判断水产动物健康程度的重要手段。研究表明,饲料中添加不同水平对水产动物血清葡萄糖、胆固醇、甘油三酯、谷丙转氨酶和谷草转氨酶等指标均可产生显著差异。Yaghoubi等[53]研究发现,当酪氨酸水平为0.85%时,苯丙氨酸缺乏可引起体重为4.7 g的银黑鲷血浆补体C3、补体C4和溶菌酶活性降低,同时引起总免疫球蛋白和血红细胞总数显著降低,而对其血浆总蛋白无显著影响。Xiao等[51]在进行饲料中不同水平苯丙氨酸(0.43%~1.91%)对杂交罗非鱼幼鱼(5.63±0.04)g血浆生化指标的影响时发现,添加苯丙氨酸可显著降低谷丙转氨酶和谷草转氨酶活性,提高血浆溶菌酶、过氧化氢酶和总蛋白活性。蒋明等[34]研究发现,吉富罗非鱼血清谷丙转氨酶含量随饲料苯丙氨酸水平的升高而先上升后下降。马志英等[41]研究表明,异育银鲫血清葡萄糖的含量在苯丙氨酸水平在1.09%~1.26%时达到最大值,这与饲喂不同苯丙氨酸水平的虹鳟[54]血清中葡萄糖变化规律一致;此外,当饲料苯丙氨酸水平为0.73%~1.67%时,异育银鲫血清甘油三酯含量先升高后下降,且在苯丙氨酸水平为1.33%时有最大值;同时,异育

银鲫血清甘油三酯含量随苯丙氨酸水平的升高而在整体上呈下降趋势，且在苯丙氨酸水平为 1.26% 和 1.58% 时出现最小值。苯丙氨酸能够调节水产动物血清生理生化的可能原因是：当水产动物体内氨基酸维持平衡后，可充分进行蛋白质的合成，此时动物体的氧化供能则主要依靠调动总胆固醇和甘油三酯等能量物质[34]。此外，当饲料中苯丙氨酸水平超出维持生长所需的需要量时，过量的氨基酸可直接参与代谢供能，也可转化为糖类或脂肪储存供能[41]。

四、苯丙氨酸对水产动物肠道健康的影响

水产动物的肠道消化酶活性决定了其对营养物质的吸收能力。研究表明，饲料中适宜水平的苯丙氨酸能提高肠消化酶的活性，促进饲料中营养物质的消化和吸收。随着饲料中苯丙氨酸水平的上升，吉富罗非鱼肠蛋白酶、肠脂肪酶活性呈先升高后下降的趋势，且在苯丙氨酸水平为 1.51% 时出现最大值，而肠淀粉酶活性变化则不显著。此外，苯丙氨酸还可改善草鱼和团头鲂溶菌酶、酸性磷酸酶和补体 C3 含量[55]。Li 等[15] 研究表明，饲料苯丙氨酸水平为 1.15% 时，草鱼肠道谷胱甘肽含量和谷胱甘肽还原酶活性有出现大值。苯丙氨酸促进水产动物肠道消化酶含量可能是通过促进胆囊收缩素的分泌并调节胰腺消化酶的分泌实现的[56]。Feng 等[57] 评估了饲料苯丙氨酸对草鱼肠道免疫反应、紧密连接蛋白基因表达水平和抗氧化相关信号通路基因表达的影响。结果表明，苯丙氨酸缺乏或过量均降低了肠道溶菌酶、酸性磷酸酶活性和补体 C3 含量。当饲料苯丙氨酸水平为 1.15% 时，可显著上调 $occludin$-1、ZO-1 和 $claudin$-c 等紧密连接蛋白的基因表达水平，同时还可显著降低白细胞介素-8（IL-8）、肿瘤坏死因子 α（TNF-α）和核因子-κB p65（NF-κB p65）的基因表达含量。此外，在苯丙氨酸水平为 1.68% 时，草鱼肠道超氧化物歧化酶（SOD）的基因表达水平显著降低；当苯丙氨酸水平为 0.91% 时，草鱼肠道核因子 E2 相关因子 2（$Nrf2$）基因表达水平显著增加。综上所述，苯丙氨酸改善鱼类的肠道免疫状态可能是通过调控雷帕霉素靶蛋白（mTOR）、NF-κB 和 Nrf2 信号通路等途径实现的[55]。

第六节　小结与展望

水产养殖业绿色发展是渔业可持续发展的必然选择。饲料中添加适宜的苯丙氨酸，不仅可以促进水产动物的生长性能，还可改善水产动物的免疫健康。目前对水产动物苯丙氨酸的研究主要集中在其需要量、血清生理生化、消化酶活性和肠道健康等几个方面，且研究对象主要集中在小规格上。然而，由于我国水产养殖存在地理分布广泛、养殖模式丰富、设施类型复杂和养殖种类多样等特殊性，水产动物在不同生长阶段的苯丙氨酸营养需要数据库还需要进一步完善。此外，关于苯丙氨酸的消化吸收、转运、代谢和与其他氨基酸相互作用的分子机制的研究还需不断深入，这不仅可以为水产动物氨基酸的利用提供参考，还可进一步推动我国水产养殖业的绿色健康发展。

参考文献

[1] Wang H C, Wang C M, Yuan W W, et al. The role of phenylalanine hydroxylase in lipogenesis in the

oleaginous fungus *Mortierella alpina* [J]. Microbiology - SGM, 2021, 167 (8): 1062.

[2] Berguig G Y, Martin N T, Creer A Y, et al. Of mice and men: plasma phenylalanine reduction in PKU corrects neurotransmitter pathways in the brain [J]. Molecular Genetics and Metabolism, 2019, 128 (4): 422 - 430.

[3] Perkowski M C, Warpeha K M. Phenylalanine roles in the seed - to - seedling stage: not just an amino acid [J]. Plant Science, 2019, 289: 110223.

[4] Gera L, Chan D C, Taraseviciene - Stewart L, et al. Rational design of small molecules for a novel class of anti - cancer drugs using a phenylalanine library [J]. Biopolymers, 2015, 80 (4): 564.

[5] Pandey S, Sunny A, Govindaraju K, et al. Development of low phenylalanine flour for phenylketonuria patient [J]. Journal of Food Processing and Preservation, 2020, 44 (11): e14894.

[6] Rudolph S, Riedel E, Henle T. Studies on the interaction of the aromatic amino acids tryptophan, tyrosine and phenylalanine as well as tryptophan - containing dipeptides with cyclodextrins [J]. European Food Research and Technology, 2018, 244 (9): 1511 - 1519.

[7] Hellwig M. Analysis of protein oxidation in food and feed products [J]. Journal of Agricultural and Food Chemistry, 2020, 68 (46): 12870 - 12885.

[8] Ghosh S, Chisti Y, Banerjee U C. Production of shikimic acid [J]. Biotechnology Advances, 2012, 30 (6): 1425 - 1431.

[9] de Oliveira M D, Araujo J D O, Galucio J M P, et al. Targeting shikimate pathway: *in silico* analysis of phosphoenolpyruvate derivatives as inhibitors of EPSP synthase and DAHP synthase [J]. Journal of Molecular Graphics & Modelling, 2020, 101: 107735.

[10] Loo Y H, Ritman P. New metabolites of phenylalanine [J]. Nature, 1964, 203 (4951): 1237 - 1239.

[11] Kawatra A, Dhankhar R, Mohanty A, et al. Biomedical applications of microbial phenylalanine ammonia lyase: current status and future prospects [J]. Biochimie, 2020, 177: 142 - 152.

[12] van Spronsen F J, Blau N, Harding C, et al. Phenylketonuria [J]. Nature Reviews Disease Primers, 2021, 7 (1): 35.

[13] Levy H L, Sarkissian C N, Scriver C R. Phenylalanine ammonia lyase (PAL): from discovery to enzyme substitution therapy for phenylketonuria [J]. Molecular Genetics and Metabolism, 2018, 124 (4): 223 - 229.

[14] van Spronsen F J. Phenylketonuria: a 21 (st) century perspective [J]. Nature Reviews Endocrinology, 2010, 6 (9): 509 - 514.

[15] Li W, Feng L, Liu Y, et al. Effects of dietary phenylalanine on growth, digestive and brush border enzyme activities and antioxidant capacity in the hepatopancreas and intestine of young grass carp (*Ctenopharyngodon idella*) [J]. Aquaculture Nutrition, 2015, 21 (6): 913 - 925.

[16] 孙丽慧, 陈建明, 潘茜, 等. 草鱼鱼种对饲料中苯丙氨酸需要量的研究 [J]. 上海海洋大学学报, 2016, 25 (3): 388 - 395.

[17] Do Nascimento T M T, Mansano C F M, Peres H, et al. Determination of the optimum dietary essential amino acid profile for growing phase of Nile tilapia by deletion method [J]. Aquaculture, 2020, 523.

[18] Zehra S, Yousif R A. Dietary total aromatic amino acid requirement and tyrosine replacement value for phenylalanine for fingerling *Oreochromis niloticus* (Linnaeus) [J]. Aquaculture Nutrition, 2021, 27 (4): 1009 - 1018.

[19] Khan M A, Abidi S F. Total aromatic amino acid requirement of Indian major carp *Labeo rohita*

(Hamilton) fry [J]. Aquaculture, 2007, 267 (1-4SI): 111-118.

[20] Benakappa S, Varghese T J. Dietary requirement of tryptophan for growth and survival of the Indian major carp, *Cirrhinus mrigala* (Hamilton-Buchanan) fry [J]. Indian Journal of Experimental Biology, 2003, 41 (11): 1342-1345.

[21] Borlongan I G, Coloso R M. Requirements of juvenile milkfish (*Chanos chanos* Forsskal) for essential amino acids [J]. The Journal of Nutrition, 1993, 123 (1): 125-132.

[22] Borlongan I G. Dietary requirement of milkfish (*Chanos chanos* Forsskal) juveniles for total aromatic amino acids [J]. Aquaculture, 1992, 102 (4): 309-317.

[23] 许琪娅, 张明亮, 瞿倩, 等. 日本鳗鲡饲料中必需氨基酸需要量的初步研究 [J]. 饲料研究, 2019, 42 (12): 55-57.

[24] Kim S S, Rahimnejad S, Song J W, et al. Comparison of growth performance and whole-body amino acid composition in red seabream (*Pagrus major*) fed free or dipeptide form of phenylalanine [J]. Asian-Australasian Journal of Animal Sciences, 2012, 25 (8): 1138-1144.

[25] Baker D H. Problems and pitfalls in animal experiments designed to establish dietary requirements for essential nutrients [J]. The Journal of Nutrition, 1986, 116 (12): 2339-2349.

[26] Robinson E H, Wilson R P, Poe W E. Total aromatic amino acid requirement, phenylalanine requirement and tyrosine replacement value for fingerling channel catfish [J]. The Journal of Nutrition, 1980, 110 (9): 1805-1812.

[27] 许红, 王常安, 徐奇友, 等. 大鳞鲃幼鱼氨基酸需要量 [J]. 华中农业大学学报, 2013, 32 (6): 126-131.

[28] Chance R E, Mertz E T, Halver J E. Nutrition of salmonoid fishes. Ⅻ. isoleucine, leucine, valine and phenylalanine requirements of chinook salmon and inter-relations between isoleucine and leucine for growth [J]. The Journal of Nutrition, 1964, 83: 177-185.

[29] Kaushik S J. Whole body amino acid composition of European seabass (*Dicentrarchus labrax*), gilthead seabream (*Sparus aurata*) and turbot (*Psetta maxima*) with an estimation of their IAA requirement profiles [J]. Aquatic Living Resources, 1998, 11 (5): 355-358.

[30] Rollin X, Mambrini M, Abboudi T, et al. The optimum dietary indispensable amino acid pattern for growing Atlantic salmon (*Salmo salar* L.) fry [J]. British Journal of Nutrition, 2003, 90 (5): 865-876.

[31] Xu Q Y, Yang J L, Wang C A. Estimating the Requirements of dietary essential amino acid pattern for young Amur Sturgeon *Acipenser schrenckii* [C] //Program & Abstracts of the 14th International Symposium on Fish Nutrition & Feeding. Qingdao: Ocean Universtiy of China, 2010: 414.

[32] Ng W K, Hung S S O. Estimating the ideal dietary indispensable amino acid pattern for growth of white sturgeon, *Acipenser transmontanus* (Richardson) [J]. Aquaculture Nutrition, 2015, 1 (2): 85-94.

[33] Kim K. Requirement for phenylalanine and replacement value of tyrosine for phenylalanine in rainbow trout (*Oncorhynchus mykiss*) [J]. Aquaculture, 1993, 113 (3): 243-250.

[34] 蒋明, 武文一, 文华, 等. 吉富罗非鱼对饲料中苯丙氨酸的需要量 [J]. 中国水产科学, 2016, 23 (5): 1173-1184.

[35] 何英霞, 许琪娅, 马德英, 等. 美洲鳗鲡必需氨基酸需要量初步研究 [J]. 广东饲料, 2019, 28 (1): 36-38.

[36] Santiago C B, Lovell R T. Amino acid requirements for growth of Nile tilapia [J]. The Journal of

Nutrition, 1988, 118 (12): 1540-1546.

[37] Forster I, Ogata H Y. Lysine requirement of juvenile *Japanese flounder Paralichthys olivaceus* and juvenile red sea bream *Pagrus major* [J]. Aquaculture, 1998, 161 (1): 131-142.

[38] Teshima S, Alam M S, Koshio S, et al. Assessment of requirement values for essential amino acids in the prawn, *Marsupenaeus japonicus* (Bate) [J]. Aquaculture Research, 2002, 33 (6): 395-402.

[39] Ren M C, Liu B, Habte-Tsion H, et al. Dietary phenylalanine requirement and tyrosine replacement value for phenylalanine of juvenile blunt snout bream, *Megalobrama amblycephala* [J]. Aquaculture, 2015, 442: 51-57.

[40] 尹东鹏,陈秀梅,刘丹妮,等,2018.乌鳢饲料赖氨酸及其他必需氨基酸营养需要量的研究 [J]. 饲料工业, 39 (14): 18-23.

[41] 马志英,朱晓鸣,解绶启,等. 异育银鲫幼鱼对饲料苯丙氨酸需求的研究 [J]. 水生生物学报, 2010, 34 (5): 1012-1021.

[42] Ngamsnae D S, Gunasekera. Arginine and phenylalanine requirement of juvenile silver perch *Bidyanus bidyanus* and validation of the use of body amino acid composition for estimating individual amino acid requirements [J]. Aquaculture Nutrition, 1999, 5 (3): 173-180.

[43] Ravi J, Devaraj K V. Quantitative essential amino acid requirements for growth of catla, *Catla catla* (Hamilton) [J]. Aquaculture, 1991, 96 (3): 281-291.

[44] Kaushik S J, Seiliez I. Protein and amino acid nutrition and metabolism in fish: current knowledge and future needs [J]. Aquaculture Research, 2010, 41 (3): 322-332.

[45] Saavedra M, Conceicao L E C, Barr Y, et al. Tyrosine and phenylalanine supplementation on *Diplodus sargus* larvae: effect on growth and quality [J]. Aquaculture Research, 2010, 41 (10): 1523-1532.

[46] Benakappa S, Varghese T J. Total aromatic amino acid requirement of the Indian major carp, *Cirrhinus mrigala* (Hamilton-Buchanan) [J]. Israeli Journal of Aquaculture-Bamidgeh, 2004, 56 (2): 131-137.

[47] Zehra S, Khan M A. Dietary phenylalanine requirement and tyrosine replacement value for phenylalanine for fingerling *Catla catla* (Hamilton) [J]. Aquaculture, 2014, 433: 256-265.

[48] Ahmed I. Dietary total aromatic amino acid requirement and tyrosine replacement value for phenylalanine in Indian major carp: *Cirrhinus mrigala* (Hamilton) fingerlings [J]. Journal of Applied Ichthyology, 2009, 25 (6): 719-727.

[49] Okrasa K, Guibe-Jampel E, Plenkiewicz J, et al. In vitro bi-enzymatic synthesis of benzaldehyde from phenylalanine: practical and mechanistic studies [J]. Journal of Molecular Catalysis B-Enzymatic, 2004, 31 (4-6): 97-101.

[50] Wyse A T S, Dos Santos T M, Seminotti B, et al. Insights from animal models on the pathophysiology of hyperphenylalaninemia: role of mitochondrial dysfunction, oxidative stress and inflammation [J]. Molecular Neurobiology, 2021, 58 (6): 2897-2909.

[51] Xiao W, Zou Z Y, Li D Y, et al. Effect of dietary phenylalanine level on growth performance, body composition, and biochemical parameters in plasma of juvenile hybrid tilapia, *Oreochromis niloticus* × *Oreochromis aureus* [J]. Journal of the World Aquaculture Society, 2020, 51 (2): 437-451.

[52] 文华,高文,罗莉,等. 草鱼幼鱼的饲料苏氨酸需要量 [J]. 中国水产科学, 2009, 16 (2): 238-247.

[53] Yaghoubi M, Mozanzadeh M T, Marammazi J G, et al. Effects of dietary essential amino acid deficiencies on the growth performance and humoral immune response in silvery-black porgy

(*Sparidentex hasta*) juveniles [J]. Aquaculture Research, 2017, 48 (10): 5311-5323.

[54] Kim K, Grimshaw T W, Kayes T B, et al. Effect of fasting or feeding diets containing different levels of protein or amino acids on the activities of the liver amino acid-degrading enzymes and amino acid oxidation in rainbow trout (*Oncorhynchus mykiss*) [J]. Aquaculture, 1992, 7 (1): 89-105.

[55] Habte-Tsion H. A review on fish immuno-nutritional response to indispensable amino acids in relation to TOR, NF-κB and Nrf2 signaling pathways: trends and prospects [J]. Comparative Biochemistry and Physiology B-Biochemistry & Molecular Biology, 2020, 241.

[56] 曾婷. 苯丙氨酸对幼建鲤消化吸收功能、抗氧化能力和免疫功能的影响 [D]. 雅安: 四川农业大学, 2011.

[57] Feng L, Li W, Liu Y, et al. Dietary phenylalanine-improved intestinal barrier health in young grass carp (*Ctenopharyngodon idella*) is associated with increased immune status and regulated gene expression of cytokines, tight junction proteins, antioxidant enzymes and related signalling molecules [J]. Fish & Shellfish Immunology, 2015, 45 (2): 495-509.

第十二章 组 氨 酸

组氨酸（Histidine，His）属于碱性氨基酸。组氨酸在水产动物的生长和健康方面发挥着重要作用，由于机体自身不能合成必须通过摄食获取来满足自身需求，属于维持水产动物生命活动的必需氨基酸之一[1]。本章综述了水产动物对组氨酸的需要量，阐明了组氨酸对水产动物肉质、抗氧化能力、消化吸收能力、免疫功能与健康的影响，讨论了组氨酸与其他氨基酸的相互作用，以期为组氨酸在水产动物营养中的应用与研究提供一定参考。

第一节 组氨酸的理化性质

组氨酸，化学名为 α-氨基 β-咪唑基丙酸，属于碱性氨基酸或杂环氨基酸。无色片状或针状结晶，无臭，稍有苦味；227 ℃软化，277 ℃分解；溶于水，极微溶于醇，不溶于乙醚。在水中旋光度−39.4°（c=1.13）。与其他氨基酸相比，除一些常见的化学反应外，由于其右侧链咪唑基与重氮苯磺酸能形成棕红色化合物，即波利（Pauly）反应。由于咪唑基解离常数为 6.0，即解离的质子浓度与水的相近，因此组氨酸既可作为质子供体，又可作为质子受体。

第二节 水产动物组氨酸需要量

组氨酸是水产动物正常生长和发育的必需氨基酸，当组氨酸过量或缺乏时会使水产动物的生长受到严重的抑制。例如，鲈组氨酸缺乏时会严重抑制其日增重，而超过鲈所需组氨酸的范围时也会严重抑制鲈日增重，说明适宜的组氨酸水平对水产动物生长发育至关重要[2]。根据表 12-1 可知，以饲料百分比计算，鱼类对组氨酸的需要量为 0.37%～1.56%，虾类在 0.8%左右。同一品种水产动物在不同发育阶段对组氨酸需要量的变化情况存在差异，李燕[3]、李燕等[2]和窦秀丽[4]的研究中分别发现，8.0 g、8.3 g 和 174.2 g 鲈的组氨酸最适需要量为 0.54%～0.57%，而 331 g 鲈的组氨酸最适需要量为 0.76%～0.78%。瞿彪[5]和 Li 等[6]在研究中分别发现，3.68 g 草鱼的组氨酸最适需要量为 1.21%，而 250 g 草鱼的组氨酸最适需要量为 0.76%～0.86%。Santiago 和 Lovell[7]、Michelato 等[8]在研究中分别发现，0.075 g 尼罗罗非鱼的组氨酸最适需要量为 0.48%，而 4.84 g 尼罗罗非鱼的组氨酸最适需要量为 0.82%。Farhat 和 Khan[9]、Khan 和 Abidi[10]的研究中分别发现，5.26 g 印度囊鳃鲇的组氨酸最适需要量为 1.51%～1.56%，而 6.6 g 印度囊鳃鲇的组氨酸最适需要量为 0.94%。Hossain 等[11]的研究中发现，3.85 g 虹鳟的组氨酸最适需要量为 0.89%，而 39.8 g 虹鳟的组氨酸最适需要量为 1.33%。Murthy 和 Varghese[12]、Ahmed 和 Khan[13]的研究中分别发现，0.61 g 印度鲮的组氨酸最适需要量为 0.85%~0.94%，而 1.88 g 露斯塔野鲮的组氨酸最适需要量为 0.90%。上述研究说明，体重和发育阶段对于水产动物组氨酸需要量存在显著的影响，且规律尚未清晰。另外，不同种类水产动物的组氨酸的需要量存在显著差异，如斑点叉尾鮰组氨酸需要量仅为 0.37%[14]，而印度囊鳃鲇组氨酸需要量高达 1.56%[10]。不同的计算模型对于氨基酸的需要量也有一定的影响，使用虚线模型和二次多项式模型对尼罗罗非鱼的组氨酸需要量进行分析，得出的需要量具有显著的差异[7,8]。针对不同指标分析得到的组氨酸的需要量存在一定差异性，选用草鱼增重率指标的组氨酸需要量为 0.76%，而选用剪切力指标的组氨酸需要量则为 0.84%[5]。但是在 Hossain 等[11]的研究中发现，饲料中添加不同水平的组氨酸对虹鳟的生长和饲料利用无显著影响，统计分析使用线性或二次多项式模型也没有显著差异。

表 12-1 水产动物组氨酸需要量

品种	体重（g）	模型	指标	需要量（占饲料干物质,%）	参考文献
鲈	8.3	二次多项式模型	特定生长率	0.57	李燕等[2]
	174.2	二次多项式模型	特定生长率	0.57	
		二次多项式模型	饲料效率	0.57	
	331.0	三次多项式模型	特定生长率	0.78	窦秀丽[4]
		三次多项式模型	饲料效率	0.76	
草鱼	8.0	二次多项式模型	特定生长率	0.54	李燕[3]
	280	二次多项式模型	增重率	0.76	瞿彪[5]
		二次多项式模型	剪切力	0.84	
		二次多项式模型	谷胱甘肽过氧化物酶	0.86	
	3.68	二次多项式模型	特定生长率	1.21	Li 等[6]
南亚野鲮	0.4	二次多项式模型	增重率	0.90	Khan 和 Abidi[15]
		二次多项式模型	饲料转化率	0.82	
建鲤	8.76	二次多项式模型	特定生长率	0.78	Zhao 等[16]
斑点叉尾鮰	200	折线模型	增重率	0.37	Wilson 等[14]
狗鲑	1.57	折线模型	增重率	0.70	Akiyama 等[17]
卡特拉鲃	1.2	虚线模型	特定生长率	0.98	Ravi 等[18]
印度鲇	4.65	二次多项式模型	特定生长率	0.54	Ahmed[19]
非洲鲇	0.22	折线回归模型	增重率	0.40	Khan 和 Abidi[20]
		二次多项式模型	饲养效率	0.41	
		二次多项式模型	蛋白质保留率	0.40	
大黄鱼	6.0	二次多项式模型	特定生长率	0.78	李燕[3]
大鳞大麻哈鱼	2.92	折线模型	增重率	0.70	Klein 和 Halver[21]
遮目鱼	0.70	折线模型	增重率	0.80	Borlongan 和 Coloso[22]

(续)

品种	体重（g）	模型	指标	需要量（占饲料干物质,%）	参考文献
尼罗罗非鱼	0.075	虚线模型	增重率	0.48	Santiago 和 Lovell[7]
	4.84	二次多项式模型	增重率	0.82	Michelato 等[8]
红姑鱼	0.98	二次折线模型	增重率	0.59	Peachey 等[23]
	6.6	二次多项式模型	增重率	0.94	Farhat 和 Khan[9]
印度囊鳃鲇	5.26	二次多项式模型	增重率	1.52	Khan 和 Abidi[10]
		二次多项式模型	蛋白质保留率	1.51	
		二次多项式模型	血红蛋白	1.56	
		二次多项式模型	红细胞比容	1.55	
团头鲂	23.30	二次多项式模型	增重率	1.12	Wilson 等[24]
		二次多项式模型	特定生长率	1.12	
卡特拉鲃	0.64	折线回归模型	增重率	0.65	Zehra 和 Khan[25]
		折线回归模型	蛋白质增重	0.64	
		折线回归模型	组氨酸增重	0.63	
		折线回归模型	血红蛋白	0.66	
虹鳟	3.85	二次多项式模型	采食量	0.89	Hossain 等[11]
	39.8	二次多项式模型	采食量	1.33	
露斯塔野鲮	1.88	折线回归模型	增重率	0.90	Murthy 和 Varghese[12]
印度鲮	0.61	二次多项式模型	增重率	0.94	Ahmed 和 Khan[13]
		二次多项式模型	饲料转化率	0.86	
		二次多项式模型	蛋白质效率	0.85	
斑节对虾	0.02	二次多项式模型	增重率	0.80	Millamena 等[26]

第三节 组氨酸对水产动物肉质的影响

鱼的增重主要是肌肉的增长，肌肉是鱼可食用的主要部分[27]。水产动物的肉质需要考虑多方面的因素，是肉类的适口性、营养价值和外观等的综合表现，而肌肉的pH、失水率、系水力和肌纤维性状等也是评价肌肉品质的重要指标[28,29]。Pei 等[30]的研究表明，饲料中缺乏组氨酸会导致草鱼肌肉中超氧化物歧化酶（SOD）、过氧化氢酶（CAT）、谷胱甘肽过氧化物酶（GPx）活性降低，肌肉中谷胱甘肽（GSH）、蛋白质和脂肪的含量显著降低，从而导致草鱼肉质的降低。在瞿彪[5]的研究中发现，饲料中添加适量组氨酸可提高草鱼肌肉中的粗蛋白和粗脂肪含量，提高肌肉的弹性、pH 和抗氧化能力，降低蒸煮损失。也有研究发现饲料中添加组氨酸可以显著提高肌肉的 pH，进而改善大西洋鲑的肉质[31]。Richter 等[32]的研究中发现，饲料中同时添加赖氨酸和组氨酸显著提高尼罗罗非鱼肌肉中蛋白质和脂肪的含量。Saavedra 等[33]的研究中发现组氨酸可调节大西洋白姑鱼的肌纤维发育，降低肌纤维面积，提高肌纤维密度。

第四节 组氨酸对水产动物抗氧化能力的影响

鱼类的组织中含有大量的不饱和脂肪酸，极易受到氧自由基的攻击而形成脂质过氧化，过氧化会改变细胞膜的流动性和通透性，导致鱼类机体氧化损伤[34]。单线态氧属于活性氧中的一种，其广泛存在于生物体内且具有很强的生物学活性，能迅速与氨基酸和不饱和脂类发生反应，氧化损伤生物大分子，甚至导致细胞凋亡[35,36]。而 Matheson 和 Lee[37]的研究中发现，组氨酸可以有效清除单线态氧。生物体内还有一种羟基自由基，是最具细胞毒性的自由基，也是造成脂质过氧化和蛋白质氧化损伤的直接原因[38]。而组氨酸是最高效的羟基自由基清除剂之一[39]，能够抑制 Fe^{3+} 诱导的比目鱼肌质网脂质过氧化[40]。组氨酸还有三种具有显著抗氧化功能的衍生二肽，分别是肌肽、鹅肌肽和高肌肽，它们能直接清除超氧阴离子、羟基自由基和过氧化氢等自由基，保护动物肌肉，使其免受自由基所造成的氧化损伤[41,42,43]。组氨酸还能提高非酶性抗氧化物质 GSH 的含量[44]，GSH 是组织细胞抵抗活性氧损害的主要低分子质量物质[45]。组氨酸也可提高血清和肌肉中谷胱甘肽还原酶（GR）的活力，从而促进 GSH 再生，增加 GSH 的含量减少机体的氧化损伤[44]。此外，在草鱼的研究中发现，组氨酸缺乏会导致抗氧化酶活性降低，包括 SOD、CAT、GPx、谷胱甘肽-S-转移酶（GST）等，而抗氧化酶的活性降低可能与核转录因子 E2 相关因子 2（Nrf2）的 mRNA 水平下降有关[46]。Nrf2 作为重要的转录因子可以调节抗氧化酶基因的表达[47]，说明 Nrf2 信号通路在组氨酸调控机体抗氧化功能中可能具有重要作用。

第五节 组氨酸对水产动物消化吸收能力的影响

水产动物的生长发育取决于对于营养物质的消化吸收能力，并且与肠道消化酶的活性密切相关[48]。鱼类的消化酶主要包括蛋白酶类、脂肪酶类和淀粉酶类。胰蛋白酶在生物体蛋白质的消化吸收过程中具有重要作用，胰蛋白酶属于丝氨酸蛋白酶家族，丝氨酸蛋白酶家族的活性中心是由丝氨酸、组氨酸和天冬氨酸残基组成的催化三联体，组氨酸在其中起到了桥梁作用[49]。在赵波[44]的研究中发现适量的组氨酸可以提高建鲤肠道中胰蛋白酶、糜蛋白酶和淀粉酶的活性。酶类是由胰脏分泌的酶原形式分泌进入肠腔，经过激活后发挥作用[50]，所以胰脏分泌消化酶对于鲤和草鱼等无胃动物具有重要作用[6,16]。在建鲤的饲料中添加组氨酸可以显著增强胰脏的谷氨酸丙酮酸转氨酶活性、胰蛋白酶、糜蛋白酶、淀粉酶、脂肪酶活性[6]，且与肠腔的酶显著相关[44]。从大比目鱼的研究中发现，水产动物消化能力与肝胰脏的生长发育情况密切相关[51]。在赵波[44]的研究中发现，组氨酸提高了建鲤肝胰脏的重量，表明组氨酸具有促进肝胰脏生长的作用，可能与提高消化能力有关。碱性磷酸酶（AKP）分布于肠道上皮刷状缘细胞表面，与多种营养物质吸收有关[52]。肌酸激酶（CK）为肠道吸收营养物质提供能量[53]。在赵波[44]的研究中发现，组氨酸能够提高建鲤肠道中 AKP 和 CK 的活力，表明组氨酸能够增强肠道吸收营养物质的能力。另外，在 Glover 等[54]的研究中发现，组氨酸可以促进虹鳟对锌的吸收。

此外，在朱强胜等[55]和邹家明等[56]的研究中发现，在翘嘴鳜脑中注射组氨酸会降低翘嘴鳜的摄食量，随着翘嘴鳜摄食量的下降氨基酸调控阻遏蛋白 2（general amino acid control non-derepressible 2，GCN2）信号通路下游激活转录因子 4（activating transcriptional factor 4，ATF4）和下丘脑中食欲因子神经肽 Y 基因（NPY）的 mRNA 表达水平显著降低，表明翘嘴鳜摄食行为的变化与 GCN2 信号通路有关；而饲料中组氨酸不足，则是通过下调促食欲因子 NPY 基因的表达对翘嘴鳜起到抑食作用。

第六节 组氨酸对水产动物免疫功能与健康的影响

赵波[44]较为系统地研究了组氨酸对于建鲤免疫能力的影响，并且在研究中发现，组氨酸可以通过调节水产动物白细胞的吞噬率、免疫细胞识别能力、黏附病原体的能力、补体 3（C3）和免疫球蛋白 M（IgM）的含量来增强水产动物的免疫能力。红细胞可以诱导静态 B 细胞增殖分化为产生免疫球蛋白的浆细胞，在水产动物免疫功能中具有重要作用[57]。组氨酸缺乏会减少血液中红细胞数量，而适量的组氨酸可以增加红细胞的数量[6,33]。作为水产动物呼吸和氨氮排泄的主要器官，鱼鳃所形成的免疫屏障和物理屏障能够抵抗病原体的入侵[58]。缺乏组氨酸会抑制细胞的抗氧化能力，诱导细胞凋亡，产生炎症，而且组氨酸缺乏会显著降低鱼鳃中紧密连接蛋白的 mRNA 表达导致鱼鳃完整性发生变化[59]，表明组氨酸缺乏可能会影响水产动物对外界病原体的抵抗力。

水产动物的肠道屏障功能在维持肠道稳态方面发挥着重要作用，肠道屏障损伤与肠道炎症发生关系密切，各种因素诱导的鱼类肠道屏障损伤能够致使肠道炎症的发生[60]。适量的组氨酸可通过下调草鱼肠道内的促炎细胞因子白细胞介素 1β（IL-1β）、白细胞介素 8（IL-8）和肿瘤坏死因子（TNF-α）的 mRNA 水平，同时上调草鱼肠道内的抗炎细胞因子白细胞介素 10（IL-10）和转化生长因子 β（TGF-β）的 mRNA 水平，进而抑制草鱼炎症的过度反应，使受伤组织开始恢复[46]。这些结果表明，组氨酸可以通过上调水产动物的抗炎细胞因子和下调促炎细胞因子来抑制肠道炎症反应。

营养性白内障主要见于鲑，与核黄素、蛋氨酸和色氨酸有关[61]，主要由于低营养浓度引起。补充结晶组氨酸可降低严重白内障的发病率[62]。在 Taylor 等[63]的研究中发现，大西洋鲑从淡水转移到海水中一年后，会有一定概率患有白内障，而在白内障早期或是发病以前补充组氨酸，则很大程度上可以预防白内障的发生。而在 Sambraus 等[64]和 Han 等[65]的研究中发现，三倍体的大西洋鲑对于组氨酸的需要量大于二倍体的大西洋鲑，且在水温升高时，三倍体大西洋鲑白内障的发病率大于二倍体大西洋鲑。据以上研究可知，组氨酸具有预防和治疗大西洋鲑白内障的功能。

第七节 组氨酸和其他氨基酸的相互作用

组氨酸与其他氨基酸存在一定相互作用，其研究主要集中在组氨酸与赖氨酸、精氨酸和脯氨酸上。较低的精氨酸补充量会显著降低日本比目鱼的生长性能，而饲料中同时添加低量的精氨酸和组氨酸会显著升高日本比目鱼的生长性能。但是饲喂添加精氨酸和组氨酸饲料的日本比目鱼会表现出较高的氧化应激和较低的溶菌酶活性[66]。与单独添加组氨酸的饲料相比，在组氨酸饲料中添加脯氨酸会显著降低食物摄入量和对体脂的积累[67]。由上述研究可知，氨基酸之间互作对于水产动物而言并非全是优点，也有缺陷。深入研究氨基酸之间的相互作用对水产动物的健康养殖具有重要意义。

第八节　小结与展望

近年来，研究人员开展了大量组氨酸的相关研究，确定了鱼类对组氨酸的需要量为0.37%~1.56%，并且较为系统的研究了组氨酸过量或缺乏对水产动物肉质、抗氧化能力、消化吸收和免疫功能的影响。但是，关于虾蟹类对组氨酸需要量的研究相对较少，因此无法确定虾蟹类对组氨酸需要量的范围。关于虾蟹类组氨酸最适需要量应成为未来研究的方向之一。未来对于组氨酸的研究，不应局限于水产动物的幼体阶段，应拓展至中后期阶段；此外，组氨酸与其他氨基酸和营养物质的互作关系也是值得深入研究的方向。这不仅为水产动物氨基酸利用提供参考，还可以进一步推动我国水产养殖行业的健康发展。

参考文献

[1] 周凡，邵庆均. 鱼类必需氨基酸营养研究进展 [J]. 饲料与畜牧，2010，7 (8)：20-26.

[2] 李燕，艾庆辉，麦康森，等. 鲈鱼对组氨酸需要量的研究 [J]. 中国海洋大学学报（自然科学版），2011，41 (3)：31-36，42.

[3] 李燕. 鲈鱼和大黄鱼支链氨基酸与组氨酸营养生理的研究 [D]. 青岛：中国海洋大学，2010.

[4] 窦秀丽. 鲈鱼（*Lateolabrax japonicus*）生长中后期缬氨酸、组氨酸和苏氨酸营养需求的研究 [D]. 大连：大连海洋大学，2014.

[5] 瞿彪. 组氨酸对生长中期草鱼肉质的影响和肠道、鳃损伤的保护作用研究 [D]. 雅安：四川农业大学，2014.

[6] Li X F, et al. Effects of graded levels of histidine on growth performance, digested enzymes activities, erythrocyte osmotic fragility and hypoxia-tolerance of juvenile grass carp *Ctenopharyngodon idella* [J]. Aquaculture, 2016, 452 (2): 388-394.

[7] Santiago C B, Lovell R T. Amino acid requirements for growth of Nile Tilapia [J]. The Journal of Nutrition, 1988, 118 (12): 1540-1546.

[8] Michelato M, Zaminhan M, Boscolo W R, et al. Dietary histidine requirement of Nile tilapia juveniles based on growth performance, expression of muscle-growth-related genes and haematological responses [J]. Aquaculture, 2016, 467 (1): 63-70.

[9] Farhat, Khan M A. Effects of varying levels of dietary L-histidine on growth, feed conversion, protein gain, histidine retention, hematological and body composition in fingerling stinging catfish *Heteropneustes fossilis* (Bloch) [J]. Aquaculture, 2013, 404 (8): 130-138.

[10] Khan M A, Abidi S F. Dietary histidine requirement of Singhi, *Heteropneustes fossilis* fry (Bloch) [J]. Aquaculture Research, 2014, 45 (8): 1341-1354.

[11] Hossain M S, Lee S, Small B C, et al. Histidine requirement of rainbow trout (*Oncorhynchus*

mykiss) fed a low fishmeal based diet for maximum growth and protein retention [J]. Aquaculture Research, 2021, 52 (2): 3785-3795.

[12] Murthy H S, Varghese T J. Arginine and histidine requirements of the Indian major carp, *Labeo rohita* (Hamilton) [J]. Aquaculture Nutrition, 2010, 1 (4): 235-239.

[13] Ahmed I, Khan M A. Dietary histidine requirement of fingerling Indian major carp, *Cirrhinus mrigala* (Hamilton) [J]. Aquaculture Nutrition, 2005, 11 (5): 359-366.

[14] Wilson R, Poe W, Robinson E, et al. Leucine, isoleucine, valine and histidine requirements of fingerling channel catfish [J]. The Journal of nutrition, 1980, 110 (4): 627-633.

[15] Khan M A, Abidi S F. Dietary histidine requirement of fingerling Indian major carp, *Labeo rohita* (Hamilton) [J]. Israeli Journal of Aquaculture-Bamidgeh, 2004, 13 (6): 424-430.

[16] Zhao B, Feng L, Liu Y, et al. Effects of dietary histidine levels on growth performance, body composition and intestinal enzymes activities of juvenile Jian carp (*Cyprinus carpio* var. Jian) [J]. Aquaculture Nutrition, 2012, 18 (2): 220-232.

[17] Akiyama T, Arai S, Murai T, et al. Threonine, histidine and lysine requirements of chum salmon fry [J]. Nippon Suisan Gakkaishi, 1985, 51 (4): 635-639.

[18] Ravi J, Devaraj K V. Quantitative essential amino acid requirements for growth of catla, *Catla catla* (Hamilton) [J]. Aquaculture, 1991, 96 (8): 281-291.

[19] Ahmed B I. Dietary amino acid l-histidine requirement of fingerling Indian catfish, *Heteropneustes fossilis* (Bloch), estimated by growth and whole body protein and fat composition [J]. Journal of Applied Ichthyology, 2013, 29 (3): 602-609.

[20] Khan M A, Abidi S F. Optimum histidine requirement of fry African catfish, *Clarias gariepinus* (Burchell) [J]. Aquaculture Research, 2009, 40 (9): 1000-1010.

[21] Klein R G, Halver J E. Nutrition of salmonoid fishes: Arginine and histidine requirements of Chinook and Coho salmon [J]. The Journal of Nutrition, 1970, 100 (9): 1105-1109.

[22] Borlongan I G, Coloso R M. Requirements of juvenile milkfish (*Chanos chanos* Forsskal) for essential amino acids [J]. The Journal of nutrition, 1993, 123 (1): 125-132.

[23] Peachey B L, Scott E M, Gatlin D M, et al. Dietary histidine requirement and physiological effects of dietary histidine deficiency in juvenile red drum *Sciaenop ocellatus* [J]. Aquaculture, 2018, 483 (1): 244-251.

[24] Wilson-Arop O M, Liang H, Ge X, et al. Dietary histidine requirement of juvenile blunt snout bream (*Megalobrama amblycephala*) [J]. Aquaculture Nutrition, 2018, 25 (1): 249-259.

[25] Zehra S, Khan M A. Dietary histidine requirement of fingerling *Catla catla* (Hamilton) based on growth, protein gain, histidine gain, RNA/DNA ratio, haematological indices and carcass composition [J]. Aquaculture Research, 2014, 47 (4): 1028-1039.

[26] Millamena O M, Teruel M B, Kanazawa A, et al. Quantitative dietary requirements of postlarval tiger shrimp, *Penaeus monodon*, for histidine, isoleucine, leucine, phenylalanine and tryptophan [J]. Aquaculture, 1999, 179 (1-4): 169-179.

[27] Fauconneau B, Alami-Durante H, Laroche M, et al. Growth and meat quality relations in carp [J]. Aquaculture, 1995, 129 (1): 265-297.

[28] 王艳萍, 曾维斌, 廖秋萍, 等. 阿克苏地区三种食用鱼肉质特性的比较分析 [J]. 淡水渔业,

2012, 42 (3): 81-83.

[29] Bickerdike, R, Hagen, et al. Effects of feed, feeding regime and growth rate on flesh quality, connective tissue and plasma hormones in farmed Atlantic salmon (*Salmo salar* L.) [J]. Aquaculture, 2011, 318 (3/4): 343-354.

[30] Pei W, Bq A, Lin F, et al. Dietary histidine deficiency induced flesh quality loss associated with changes in muscle nutritive composition, antioxidant capacity, Nrf2 and TOR signaling molecules in on-growing grass carp (*Ctenopharyngodon idella*) [J]. Aquaculture, 2020, 526 (9): 735518.

[31] Forde-Skjaervik O, Refstie S, Aslaksen M A, et al. Digestibility of diets containing different soybean meals in Atlantic cod (*Gadus morhua*): comparison of collection methods and mapping of digestibility in different sections of the gastrointestinal tract [J]. Aquaculture, 2006, 261 (1): 241-258.

[32] Richter B L, Silva T, Michelato M, et al. Combination of lysine and histidine improves growth performance, expression of muscle growth related genes and fillet quality of grow out Nile tilapia [J]. Aquaculture Nutrition, 2020, 27 (2): 568-580.

[33] Saavedra M, Pereira T G, Ribeiro L, et al. Evaluation of dietary histidine supplementation on meagre, *Argyrosomus regius*, juvenile growth, haematological profile, stress and muscle cellularity [J]. Aquaculture Nutrition, 2020, 26 (4): 1223-1230.

[34] Frankel E N. Lipid oxidation: mechanisms, products and biological significance [J]. Journal of the American Oil Chemists Society, 1984, 61 (12): 1908-1917.

[35] Claude, Schweitzer, Reinhard, et al. Physical mechanisms of generation and deactivation of singlet oxygen [J]. ChemInform, 2003, 34 (29): 1685-1758.

[36] Khan A U, Kasha M. Singlet molecular oxygen in the Haber-Weiss reaction [J]. Proceedings of the National Academy of Sciences, 1994, 91 (26): 12365-12367.

[37] Matheson I, Lee J. Chemical reaction rates of amino acids with singlet oxygen [J]. Photochemistry and Photobiology, 2010, 29 (5): 879-881.

[38] Halliwell B, Chirico S. Lipid peroxidation: its mechanism, measurement, and significance [J]. American Journal of Clinical Nutrition, 1993, 57 (5): 715-725.

[39] Zs.-Nagy I, Floyd R A. Hydroxyl free radical reactions with amino acids and proteins studied by electron spin resonance spectroscopy and spin-trapping [J]. Biochimica et Biophysica Acta (BBA)-Protein Structure and Molecular Enzymology, 1984, 790 (3): 238-250.

[40] Erickson M C, Hultin H O. Influence of histidine on lipid peroxidation in sarcoplasmic reticulum [J]. Archives of Biochemistry & Biophysics, 1992, 292 (2): 427-432.

[41] Quinn P J, Boldyrev A A, Formazuyk V E. Carnosine: Its properties, functions and potential therapeutic applications [J]. Molecular Aspects of Medicine, 1992, 13 (5): 379-444.

[42] Begum G, Cunliffe A, Leveritt M. Physiological role of carnosine in contracting muscle [J]. International Journal of Sport Nutrition & Exercise Metabolism, 2005, 15 (5): 493-514.

[43] 韩立强, 杨国宇, 王艳玲, 等. 肌肽清除自由基及抗氧化性质的作用研究 [J]. 河南工业大学学报: 自然科学版, 2006, 27 (1): 43-46.

[44] 赵波. 组氨酸对幼建鲤消化吸收功能、抗氧化能力和免疫功能的影响 [D]. 雅安: 四川农业大学, 2011.

[45] Finkel T. Redox-dependent signal transduction [J]. FEBS Letters, 2000, 476 (1/2): 52-54.

[46] Jiang W D, Feng L, Qu B, et al. Changes in integrity of the gill during histidine deficiency or excess due to depression of cellular anti-oxidative ability, induction of apoptosis, inflammation and impair of cell-cell tight junctions related to Nrf2, TOR and NF-κB signaling in fish [J]. Fish & Shellfish Immunology, 2016, 56 (9): 111-122.

[47] Szklarz G. Role of Nrf2 in oxidative stress and toxicity [J]. Annual Review of Pharmacology and Toxicology, 2013, 53 (1): 401-426.

[48] Kolkovski S, Tandler A, Kissil G W, et al. The effect of dietary exogenous digestive enzymes on ingestion, assimilation, growth and survival of gilthead seabream (*Sparus aurata*, Sparidae, Linnaeus) larvae [J]. Fish Physiology & Biochemistry, 1993, 12 (3): 203-209.

[49] Polgár L. The catalytic triad of serine peptidases [J]. Cellular & Molecular Life Sciences Cmls, 2005, 62 (19-20): 2161-2172.

[50] Slack J. Developmental Biology of the Pancreas [J]. Development, 1995, 121 (6): 1569-1580.

[51] Gisbert E, Piedrahita R H, Conklin D E. Ontogenetic development of the digestive system in California halibut (*Paralichthys californicus*) with notes on feeding practices [J]. Aquaculture, 2004, 232 (1/2/3/4): 455-470.

[52] Villanueva J, Vanacore R, Goicoechea O, et al. Intestinal alkaline phosphatase of the fish *Cyprinus carpio*: Regional distribution and membrane association [J]. Journal of Experimental Zoology Part A Ecological Genetics & Physiology, 1997, 279 (4): 347-355.

[53] Wallimann T, Hemmer W. Creatine kinase in non-muscle tissues and cells [J]. Springer US, 1994, 13 (7): 193-220.

[54] Glover C N, Bury N R, Hogstrand C, et al. Zinc uptake across the apical membrane of freshwater rainbow trout intestine is mediated by high affinity, low affinity, and histidine-facilitated pathways [J]. Biochimica et Biophysica Acta (BBA)-Biomembranes, 2003, 1614 (2): 211-219.

[55] 朱强胜, 何珊, 梁旭方, 等. 组氨酸及组胺对翘嘴鳜摄食调控的影响 [J]. 华中农业大学学报, 2020, 39 (6): 180-186.

[56] 邹家明, 何珊, 梁旭方, 等. 脑室注射和饲料缺乏组氨酸或缬氨酸对翘嘴鳜摄食的调控作用 [J]. 华中农业大学学报(自然科学版), 2022, 41 (2): 168-175.

[57] Carroll, Michael C. The complement system in regulation of adaptive immunity [J]. Nature Immunology, 2004, 5 (10): 981-986.

[58] Habte-Tsion H M, Ge X, Liu B, et al. A deficiency or an excess of dietary threonine level affects weight gain, enzyme activity, immune response and immune-related gene expression in juvenile blunt snout bream (*Megalobrama amblycephala*) [J]. Fish & Shellfish Immunology, 2015, 42 (2): 439-446.

[59] 陈秀梅, 王桂芹, 单晓枫, 等. 鱼类肠道屏障损伤与肠道炎症发生发展关系的研究进展 [J]. 河南农业科学, 2022, 51 (5): 1-9.

[60] 李岩. 浅谈鱼类营养性疾病 [J]. 国外水产, 1989, (1): 19-23.

[61] Breck O, Bjerkå S E, Campbell P, et al. Cataract preventative role of mammalian blood meal, histidine, iron and zinc in diets for Atlantic salmon (*Salmo salar* L.) of different strains [J]. Aquaculture Nutrition, 2015, 9 (5): 341-350.

[62] Waagbo R, Trosse C, Koppe W, et al. Dietary histidine supplementation prevents cataract development in

adult Atlantic salmon, *Salmo salar* L., in seawater [J]. British Journal of Nutrition, 2010, 104 (10): 1460-1470.

[63] Taylor, R, et al. Adult triploid Atlantic salmon (*Salmo salar*) have higher dietary histidine requirements to prevent cataract development in seawater [J]. Aquaculture Nutrition, 2014, 21 (1): 18-32.

[64] Sambraus F, Fjelldal P G, Remo S C, et al. Water temperature and dietary histidine affect cataract formation in Atlantic salmon (*Salmo salar* L.) diploid and triploid yearling smolt [J]. Journal of Fish Diseases, 2017, 40 (9): 1195-1212.

[65] Han Y, Koshio S, Ishikawa M, et al. Interactive effects of dietary arginine and histidine on the performances of Japanese flounder *Paralichthys olivaceus* juveniles [J]. Aquaculture, 2013, 414: 173-182.

[66] Asahi R, Tanaka K, Fujimi T J, et al. Proline decreases the suppressive effect of histidine on food intake and fat accumulation [J]. Journal of Nutritional Science & Vitaminology, 2016, 62 (4): 277.

第三篇
其他氨基酸及肽营养

第十三章 牛 磺 酸

牛磺酸在水产动物体内常以含硫游离氨基酸或小分子二肽或三肽的形式存在，其主要生物学功能在于调节机体正常生理机能，如形成结合胆汁酸、调节渗透压、维持细胞膜稳定性及提升抗氧化能力等。在"缺乏牛磺酸的植物性蛋白源及某些新型蛋白源广泛应用于水产饲料"的背景下，水产动物牛磺酸营养研究逐步受到重视。本章综述了水产动物饲料牛磺酸的来源、水产动物牛磺酸的合成能力、水产动物牛磺酸营养需要量、参与水产动物营养物质的代谢、牛磺酸对水产动物健康的影响、牛磺酸与其他营养物质的相互作用，以期为牛磺酸在水产动物营养免疫与健康养殖方面的进一步研究提供科学理论基础。

第一节 牛磺酸的理化性质

牛磺酸（Taurine，Tau），分子式 $C_2H_7NO_3S$，分子质量 125.15，熔点 305.11 ℃，分子结构如图 13-1 所示，白色结晶粉末，无毒、无臭、味微酸、对热稳定。牛磺酸可溶于水，12 ℃时溶解度为 0.5%，其水溶液 pH 为 4.1~5.6，在 95% 乙醇中 17 ℃时溶解度为 0.004%，不溶于无水乙醇、乙醚和丙酮。

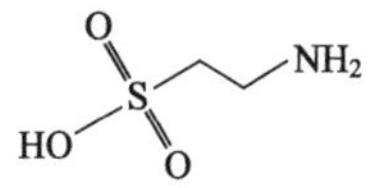

图 13-1 牛磺酸的分子结构

第二节 水产动物饲料牛磺酸的来源

现有的研究表明水产动物自身合成牛磺酸的能力相对较弱[1]，为满足自身需求则需要从饲料中获取，因此牛磺酸对于水产动物而言是一种条件性必需氨基酸[2]。在常用的水产饲料原料中，鱼粉等动物性蛋白原料为饲料配方中牛磺酸的主要提供者，而大多数植物性蛋白原料中牛磺酸含量较少或基本为零（表 13-1）[3]。由于养殖环境恶化和饲料成本升高等产业现状，鱼粉等动物性蛋白原料被越来越多的植物性蛋白源所替代，牛磺酸成为主要限制性营养物质之一。因此，在未来饲料配方研发及新型蛋白源开发的过程中，如何满足水产动物牛磺酸营养应成为主要关注点之一。

表 13-1 常见水产饲料原料的牛磺酸含量（mg/kg）

原料	牛磺酸含量	原料	牛磺酸含量
鱼粉	3 201	肉粉	1 150
虾粉	1 094	鸡肉粉	180
鱼蛋白水解物	7 501	家禽副产物粉	3 079
肉骨粉	386	酵母	112

第三节 水产动物牛磺酸合成能力

水产动物牛磺酸合成包括4种途径（图13-2），其中途径1和途径2为水产动物牛磺酸合成的主要途径。水产动物牛磺酸的合成能力主要取决于牛磺酸合成关键酶的活力。半胱亚磺酸脱羧酶（cysteine sulfinate decarboxylase，CSD）和半胱胺双加氧酶（cysteine dioxygenase，CDO）是分别为途径1和途径2的关键限速酶[4]。目前关于水产动物牛磺酸合成能力的研究主要集中于鱼类，缺乏对于虾蟹类的研究。肝脏是鱼类牛磺酸合成的主要器官，表13-2列举了主要养殖鱼类肝脏牛磺酸合成关键酶的活力[4]。研究表明，鱼的种类是影响肝脏牛磺酸合成能力的重要因素。一般来说，海水鱼类的牛磺酸合成能力低于淡水鱼类，甚至无合成能力[5]。不同鱼类肝脏牛磺酸合成途径存在差异，多数鱼类主要通过途径1合成牛磺酸，而对于鲤[1,6]、香鱼[1,9]及太阳鱼[1]来说，当途径1合成的牛磺酸无法满足自身需求时，仍可通过途径2进行合成。

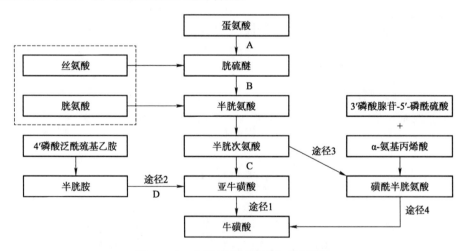

图 13-2 水产动物牛磺酸合成途径[4]

注：A代表胱硫醚合成酶，B代表胱硫醚酶，C代表半胱次磺酸脱羧酶，D代表半胱胺双加氧酶。

表 13-2 鱼类肝脏牛磺酸合成关键酶的活力（酶活力/体重，U/g）

种类	途径1	途径2
淡水鱼类		
虹鳟	0.544/(40.6)[5]	
	0.666/(306.7)[5]	

(续)

种类	途径1	途径2
鲤	0.30/(35)[1]	0.86 (72~162)[6]
	0.011/(372)[5]	
罗非鱼	0.558/(86.0)[5]	
	0.05/(2500)[1]	
香鱼	2.33/(135~156)[1]	0.09/(141167)[7]
太阳鱼	15.82/(25~40)[1]	3.06/(25~40)[7]
		1.85/(68~90)[7]
海水鱼类		
日本牙鲆	0.203/(30.1)[5]	
	0.274/(60.3)[5]	
	0.388/(126.1)[5]	
圆斑星鲽	0.068/(13.0±1.5)[5]	
	0.237/(64.1±7.5)[5]	
木叶鲽	0.05/(100)[1]	
金枪鱼	0.001/(138.4)[5]	
五条鰤	0.002/(98.6)[5]	
	0.011/(3 900)[5]	
飞鱼	0.001/(3 109)[5]	
真鲷	0.245/(1 071)[5]	2.04/(0.2~1.0 kg)[1]

第四节 水产动物牛磺酸需要量

近十余年发表的水产动物牛磺酸需要量研究见表 13-3。水产动物牛磺酸需要量受到生长阶段、评价指标、饲料中蛋白组成及养殖环境的影响。在饲料蛋白源统一的条件下，体重为 (6.87±0.12) g[11]、(38.1~38.9) g[12] 草鱼的牛磺酸需要量分别为 0.12%、0.06%；体重为 (6.3±0.01) g、(165.9±5.01) g 大菱鲆的牛磺酸需要量分别为 1.15%、0.64%[21]。在 (0.13±0.01) g 虹鳟粉饲料中添加 0.5%~1.5% 牛磺酸未能促进生长[16]，但在 (203.26±2.01) g 虹鳟粉饲料中添加 0.06% 能显著提升生长性能[18]。上述研究表明规格对于水产动物牛磺酸需要量的影响规律仍需进一步探索。在不同生长阶段，水产动物对于牛磺酸的需要量亦存在差异。以增重率为评价指标，尼罗罗非鱼幼鱼在 28 d、56 d 及 84 d 时的牛磺酸需要量分别为 0.72%、0.68% 及 0.64%[14]，在对斜带石斑鱼幼鱼[20]的研究中得到了类似的变化规律。评价指标的变化也影响着牛磺酸需要量，黄颡鱼以特定生长率为评价指标时所得出的牛磺酸需要量低于以血清溶菌酶为评价指标时[22]。鱼粉是水产动物饲料中牛磺酸的主要来源，而植物蛋白源中牛磺酸相对缺乏，因此植物蛋白源替代鱼粉时应考虑水产动物对牛磺酸的需要变化。田芊芊等[10]发现在豆粕蛋白替代50%鱼粉的饲料中，青鱼牛磺酸需要量为0.1%。在对大口黑鲈[19]、红鳍东方鲀[27]、凡纳滨对虾[32]等研究中均得到了一致的结果。不同养殖环境对水产动物牛磺酸需要量有着潜在影响，李航等[31]在淡水养殖条件下得出的（0.160±0.002）g 凡纳滨对虾牛磺酸的需要量为0.44%～0.58%。而Yue[30]发现在正常海水盐度下 0.48 g 凡纳滨对虾牛磺酸需要量为0.17%。

表 13-3 水产动物牛磺酸需要量

种类	体重（g）	主要蛋白质源	模型	指标	需要量（占饲料干物质,%）	参考文献
淡水鱼类						
鲫	29.07	鱼粉、豆粕	—	增重率	0.6	伍琴等[8]
翘嘴鲌	3.14	豆粕蛋白替代35%鱼粉	—	生长、体成分、血清生化指标	1	刘汝鹏等[9]
青鱼	5.90	豆粕蛋白替代50%鱼粉	—	提高生长性能、改善肠道结构、提高抗急性拥挤胁迫能力	0.1	田芊芊等[10]
草鱼	6.87	豆粕、菜籽粕	二次多项式	增重率	0.12	徐璐茜等[11]
草鱼	38.1～38.9	豆粕、菜籽粕	—	促进生长、提高饲料转化率和蛋白质效率	0.06	罗莉等[12]
黄颡鱼	5.18	豆粕、玉米蛋白粉、小麦蛋白粉	折线模型	增重率	1.09	Li 等[13]
				血清溶菌酶	1.60	
尼罗罗非鱼	4.25	酪蛋白	二次多项式	增重率	26 d：0.72 56 d：0.68 84 d：0.62	何凌云等[14]
吉富罗非鱼	5.89	酪蛋白	二次多项式	增重率	0.75	王和伟等[15]
虹鳟	0.13	鱼粉	—	未促进生长（0.5%～1.5%）	—	徐奇友等[16]
虹鳟	18.4	豆粕、玉米蛋白粉、小麦蛋白粉		生长、肉质比	0.85	Gaylord 等[17]
	203.26	鱼粉		提升生长性能、抗氧化能力	0.06	罗志成[18]
大口黑鲈	12.6	豆粕蛋白替代60%鱼粉		促进生长	0.5	Wu 等[19]
海水鱼类						
斜带石斑鱼	13.85	酪蛋白	二次多项式	增重率	26 d：1.20 56 d：1.08 84 d：1.00	王学习等[20]
大菱鲆	6.3	酪蛋白、明胶	—	增重率	1.15	Qi 等[21]
	165.9				0.64	

(续)

种类	体重（g）	主要蛋白质源	模型	指标	需要量（占饲料干物质,%）	参考文献
日本牙鲆	0.90	鱼粉	—	增重率	1.4	Park 等[22]
	0.3			增重率、饲料效率	1.60	Kim 等[23]
	3.7				0.56	
海鲈	0.79	鱼粉、豆粕	—	特定生长率	0.45	Martinez 等[24]
真鲷	0.5	酪蛋白、明胶	二次多项式	增重率	0.48	Matsunari 等[25]
				饲料效率	0.52	
	580	大豆浓缩蛋白	—	特定生长率、饲料效率	0.56	Takagi 等[26]
红鳍东方鲀	17.33	豆粕蛋白替代30%鱼粉	—	特定生长率	0.5	郭斌等[27]
	32.28	鱼粉、酪蛋白、豆粕	折线模型	黏液溶菌酶、肝脏总抗氧化能力、肠道脂肪酶活力	0.14, 0.18, 2.21	周婧等[28]
龙胆石斑鱼	14.42	豆粕蛋白替代40%鱼粉	—	提高生长和营养物质消化率	0.1	Lin and Lu[29]
甲壳类						
凡纳滨对虾（海水养殖）	0.48	豆粕、鱼粉	二次多项式	增重率	0.17	Yue 等[30]
凡纳滨对虾（淡水养殖）	0.160	豆粕、鱼粉	—	促进生长、改善抗亚硝酸盐和低氧胁迫能力	0.44~0.58	李航等[31]
凡纳滨对虾	1.19	豆粕蛋白替代50%鱼粉	—	促进生长、改善免疫	0.1	To and Liou[32]
克氏原螯虾	4.50	菜粕、豆粕、鱼粉	—	促进生长、提高非特异性免疫力	0.02~0.03	孔圣超等[33]

第五节　牛磺酸对水产动物营养物质代谢的影响

一、牛磺酸对脂肪代谢的影响

牛磺酸对脂肪代谢的影响情况在不同水产动物的研究中存在一定差异。在饲料中添加牛磺酸可以降低草鱼[12]、斜带石斑鱼[20]、龙胆石斑鱼[29]、杂交条纹鲈[34]体脂肪含量。牛磺酸的重要生理功能之一是与胆酸结合形成牛磺胆酸等结合胆汁酸，结合胆汁酸通过乳化脂肪，增大脂肪与脂肪酶接触面积，促进脂肪消化吸收，减少体内脂肪沉积量[35]。在纯植物蛋白源饲料中至少需添加0.33%牛磺酸才能够避免塞内加尔鳎肝脏胆汁酸盐合成的中断并维持脂质消化吸收能力[36]。此外，增强脂肪消化、分解及吸收能力是牛磺酸减少体内脂肪沉积的重要原因之一。投喂添加牛磺酸0.05%、0.1%、0.15%或0.2%饲料8周的鲤肠道脂肪酶活力均高于对照组，显著改善鲤对饲料脂肪的消化和吸收能力[37]。

王清滨[38]为了探讨高脂饲料中添加牛磺酸对草鱼幼鱼脂代谢的影响，在高脂饲料中添加0.07%~0.17%的牛磺酸，发现牛磺酸有效提高肝胰脏中脂蛋白脂酶（lipoprotein lipase，LPL）、肝脂酶和总脂酶等脂肪分解酶类的活性，增加肝脏中甘油三酯的分解，并提高胆固醇7α-羟化酶（CYP7A1）的活性及其mRNA的表达量，促进肝脏中胆固醇降解为胆汁酸，进而降低机体内脂肪含量。而对于尼罗罗非鱼[14]和凡纳滨对虾[31]饲料中添加牛磺酸却会增加鱼体脂肪含量。由此可见，牛磺酸在水产动物脂肪代谢可能发挥着关键作用，但在不同水产动物中的具体调控机制仍需深入研究。

二、牛磺酸对糖代谢的影响

牛磺酸能够缓解浅水胁迫应激引起的黄斑篮子鱼[39]和团头鲂[40]血清血糖水平升高，也可以显著降低青鱼[19]急性拥挤胁迫后的血清葡萄糖水平，表明牛磺酸能够参与水产动物葡萄糖稳态的调节。Zhang等[41]以未添加牛磺酸的15%糖水平组作为对照，采用2×3双因素（糖水平：15%、30%；牛磺酸水平：0.4%、1.2%和2.0%）试验明确了在高糖饲料条件下牛磺酸能够通过显著提升大菱鲆肝脏糖酵解、葡萄糖转运、糖原合成能力，抑制糖异生水平，维持机体正常血糖水平，且在对肝脏糖酵解及葡萄糖转运载体基因表达影响上牛磺酸和饲料糖水平存在显著的交互作用。综合以上研究可以看出，牛磺酸对水产动物的糖代谢机能的具有一定调节作用，但仍需从与激素的协同作用、改善胰岛素等角度更深入探讨牛磺酸调控水产动物糖代谢的潜在机制。

三、牛磺酸对蛋白质代谢的影响

尽管牛磺酸不参与水产动物体内蛋白质的合成，但在草鱼[12]、斜带石斑鱼[15]、凡纳滨对虾[41]、鲤[42]及中华绒螯蟹[43]的饲料中添加牛磺酸均能够提高水产动物机体蛋白质含量。对于饲料中添加适量牛磺酸增强体蛋白沉积的解释，目前可以归纳为3种观点，但具体机制仍需要进一步研究。首先，牛磺酸提升水产动物体内蛋白质的消化吸收效率是其中重要原因之一。伍琴等[8]探讨牛磺酸对鲫蛋白质消化吸收相关基因表达影响时发现，牛磺酸通过调节氨肽酶N（APN）、小肽转运蛋白（PepT1）、基因系尾型同源盒基因（CDX2）、L-氨基酸转运载体蛋白（LAT2）mRNA在鲫肠道中的相对表达量来调控蛋白质的水解、小肽的转运以及氨基酸的消化吸收，当牛磺酸含量为0.6%时上述基因表达量达到最高。其次，牛磺酸能够调节蛋白质代谢相关激素的分泌。邱小琼和赵红雪发现0.2%、0.4%、0.6%及0.8%牛磺酸添加到基础饲料中能够显著提高鲤血清甲状腺激素含量，血清甲状腺激素决定着鱼类生长激素的基因表达和合成，而生长激素能够增强鱼类蛋白质合成能力[44]。最后，体内牛磺酸的生物合成多为含硫氨基酸参与，而外源性牛磺酸的补充能够减少含硫氨基酸用于合成牛磺酸的比例，更多用于蛋白质合成，从而提升蛋白质质量，增强蛋白质的可吸收性，提高蛋白质利用率[51]。

第六节 牛磺酸对水产动物健康的影响

一、牛磺酸对水产动物抗氧化能力及免疫力的影响

目前已经发现饲料中补充适宜牛磺酸能够提升水产动物的抗氧化能力及免疫力。徐璐

茜等[11]在对（6.87±0.12）g草鱼的研究中发现，饲喂1.0 g/kg及1.5 g/kg牛磺酸能够显著提高肝脏抗氧化能力及非特异性免疫力。孔圣超等[33]在研究克氏原螯虾时发现，在饲料中添加300 mg/kg牛磺酸可在4周后显著提高肝胰脏超氧化物歧化酶（SOD）和过氧化氢酶（CAT）活力，抑制丙二醛（MDA）的产生，提升血清中溶菌酶（LZM）的活性，增强鱼体抗氧化及免疫能力。Li等[13]研究表明，当全植物蛋白饲料中牛磺酸含量达到1.09%时可以显著提高黄颡鱼SOD、CAT及谷胱甘肽过氧化物酶（GPx）活力，加强黄颡鱼清除自由基的能力，进而提升抗氧化能力。王清滨[38]研究表明，高脂饲料中添加700~1 200 mg/kg牛磺酸显著提高抗氧化酶活性，减少脂质过氧化产物产生，减轻脂质过氧化程度，缓解高脂对机体的氧化损伤。Kumar等[45]对淡水鲇研究发现，饲料中5 mg/kg牛磺酸通过降低肝脏、肾脏及红细胞的脂质过氧化物浓度，增强SOD和CAT活性，改善镉诱导的氧化应激。此外，黄岩[46]在牛磺酸改善CCl_4诱导斜带石斑鱼原代肝细胞氧化损伤的研究中发现，10 mmol/L牛磺酸不仅能够显著增强肝细胞活力及抗氧化酶活性，同时上调了CAT和$GSH-Px$的基因表达量，进而保护和修复因氧化损伤导致的鱼体抗氧化能力下降，从细胞层面初步揭示了牛磺酸抗氧化的分子机制。细胞因子介导的炎症反应在水产动物免疫应答及抗氧化中发挥着重要作用。Yan等[47]研究发现，全植物蛋白饲料中添加0.99%牛磺酸可提升草鱼幼鱼肠道中LZM、补体3（C3）、补体4（C4）、酸性磷酸酶（ACP）、免疫球蛋白M（IgM）等抗菌物质的活力和含量，上调抗菌肽（$LEAP-2A$、$LEAP-2B$）的基因表达，并通过抑制核因子-κB抑制蛋白激酶（IKKs）/核因子-κB抑制蛋白α（IκBα）/核转录因子κB（NF-κB）信号通路，下调肿瘤坏死因子α（$TNF\alpha$）、Ⅱ型干扰素γ（$IFN-\gamma$）等促炎细胞因子的基因表达，以及通过增强TOR信号途径，上调白介素1β（$IL1\beta$）等抗炎细胞因子的基因表达，缓解嗜水气单胞杆菌感染后引发的炎症反应，改善了肠道免疫屏障功能，降低肠炎发病率。

二、牛磺酸对水产动物抗胁迫应激能力的影响

水产动物应激时的生理变化与体内营养状况紧密关联。牛磺酸除了能够提升水产动物抗氧化能力及免疫力，还能够提升水产动物抗胁迫应激能力[48]。罗志成等[18]研究了基础饲料中添加牛磺酸对慢性热应激条件下虹鳟血清生化指标的影响，发现添加1 200 mg/kg牛磺酸组在24 ℃及25 ℃慢性热应激24 h后拥有较未添加组更高的血清SOD、GSH-Px、LZM的活力和白蛋白的含量，缓解了因热应激产生的氧化损伤，显著增强了机体抗氧化能力、非特异性免疫能力及抗损伤能力，揭示了牛磺酸可成为虹鳟抗热应激的潜在饲料添加剂。石立冬等[49]在研究牛磺酸对红鳍东方鲀的热应激调控的影响时发现，红鳍东方鲀的热应激调控机制会随着添加牛磺酸的含量增加而发生变化，高添加量组依赖于代谢调节与神经活性配体-受体间相互作用调控机体对温度的响应，而低添加量组则依赖于细胞黏附分子与神经活性配体-受体间相互作用调控机体对温度的响应。Zhou等[50]为了探讨牛磺酸对冷应激条件下凡纳滨对虾生长和免疫状况的影响，结果表明，当温度从30 ℃降低至16 ℃，在饲料中添加0.2%牛磺酸有效抑制活性氧自由基的产生及非特异性脂酶（NSE）的活性，并通过增加透明细胞比例、降低半颗粒细胞及颗粒细胞的比例以改善细胞黏附性，进而抑制细胞凋亡。Tan等[51]以杂交乌鳢为试验模型，评估在饲料中补充牛磺酸是否能减

轻氨氮胁迫应激。结果表明，饲料中补充 6.0～15.0 g/kg 牛磺酸能提高血清总蛋白含量及碱性磷酸酶活性、肝脏 CAT、谷胱甘肽还原酶（GR）及热应激蛋白-70（HSP-70）基因表达量，减少活性氧簇（ROS）含量、细胞凋亡比例，有效缓解氨氮胁迫产生的应激损伤，降低死亡率。此外，在对稻田养殖黄鳝的研究中发现，低鱼粉饲料中添加 1.5 g/kg 牛磺酸可降低空气暴露应激 24 h 过程中血清中肾上腺素、皮质醇及血糖浓度，有效缓解空气暴露产生的应激反应，同时也可提升血清中 SOD 及总抗氧化能力（T-AOC）活力，并抑制 MDA 的产生，有效低空气暴露引发的氧化损伤[52]。综合以上可知，牛磺酸对不同应激源造成的鱼体应激反应及损伤均有良好的改善作用，但需进一步研究确定牛磺酸针对不同种类水产动物及不同种类应激源的适宜剂量和具体调控机制。

第七节 牛磺酸与其他营养物质的相互作用

水产动物体内的牛磺酸主要由蛋氨酸等含硫氨基酸进行合成，因此为了使体内更多的蛋氨酸等含硫氨基酸用于蛋白质合成，会在饲料中补充适量蛋氨酸等含硫氨基酸用于牛磺酸合成。但由于水产动物牛磺酸合成能力较弱，会同时补充牛磺酸以满足水产动物需要，因此在水产动物上已开展了关于牛磺酸和蛋氨酸间相互作用的研究。Gibson 等[53]采用 3×3 双因素（牛磺酸和蛋氨酸含量均为：0、5 和 10 g/kg）试验明确了当饲料中没有合成蛋氨酸添加时，5 g/kg 牛磺酸组虹鳟的增重率及含肉率显著提升，而当饲料中补充蛋氨酸时各组间虹鳟的增重率及含肉率并未出现显著差异，同时发现牛磺酸和蛋氨酸会对血浆胰岛素样生长因子 1（IGF-1）、亚牛磺酸、蛋氨酸、胱氨酸及胱硫醚含量、含肉率产生显著的交互作用，表明全植物蛋白饲料中补充牛磺酸是必要的，且蛋氨酸的补充并不会影响水产动物对饲料中牛磺酸的需要，同样的结果在波斯鲟[54]及欧洲鲈[55]的研究中也得到了证实。

牛磺酸的重要生理功能之一是参与脂类物质的消化吸收，尤其是胆固醇的代谢[56]。而在植物蛋白替代鱼粉条件下，胆固醇作为水产动物必需的营养素，往往需要饲料外源胆固醇进行补充[57,58]。因此，在水产动物上也开展了关于牛磺酸和胆固醇间相互作用的研究。Yun 等[59]以大菱鲆为研究对象探讨牛磺酸和胆固醇对其生长及胆固醇代谢的影响时发现，高植物蛋白饲料中添加 1.0%胆固醇、1.0%牛磺酸或者二者的组合可显著提高大菱鲆的生长性能，同时添加牛磺酸和胆固醇有叠加效应；此外，在高胆固醇饲料基础上添加牛磺酸可通过降低大菱鲆肝脏中 3-羟基-3 甲基戊二酸单酰 CoA 还原酶（HMG-CoAr）mRNA 相对表达量，抑制肝脏中胆固醇合成，进而降低血浆中胆固醇含量。

牛磺酸在提升水产动物免疫能力和抗病方面具有重要作用，因此在水产动物上也开展了牛磺酸与相关营养物质在此方面的相互作用研究。Han 等[60]研究发现，虽然牛磺酸和谷氨酰胺在促进鱼类生长、饲料利用、抗氧化能力和免疫能力，特别是肠道发育和结构方面具有相似的功能，但二者对日本牙鲆生长性能、血液生化及抗氧化酶活性等相关指标未产生交互作用。Kals 等[61]在对欧洲鳎的研究中发现，贫血欧洲鳎的红细胞比容和血红蛋白水平受到饲料中维生素 B_{12} 的影响，高浓度的维生素 B_{12} 会增加红细胞比容和血红蛋白水平以改善欧洲鳎贫血，而牛磺酸水平的增加则会抑制这一现象，表明二者间存在颉颃作用；此外，这种颉颃作用还表现在血浆中 Zn^{2+} 的吸收效率上，当牛磺酸和维生素 B_{12} 同时处于高浓度添加时欧洲鳗血浆中 Zn^{2+} 的吸收效率显著低于其余各组。红细胞比容和血红蛋白水平的增加会刺激铜锌超氧化物歧化酶（CuZnSOD）的产生，但由于二者的颉颃作用导致 Zn^{2+} 的吸收效率下降，CuZnSOD 的产生受阻，进而影响机体免疫力。因此，饲料中添加牛磺酸应考虑其与相关营养物质的相互作用，避免产生颉颃作用，影响牛磺酸生理功能的有效发挥。

第八节 小结与展望

在植物蛋白源替代鱼粉及新型蛋白源开发的行业大背景下，饲料中氨基酸平衡问题及功能性营养物质缺乏问题凸显，不仅会造成水产动物机体产生生理病理变化，引发营养性疾病，还会增加饲料及药物使用量，提高养殖成本。而牛磺酸正是植物蛋白源及某些新型蛋白源易缺乏的功能性氨基酸之一。目前，牛磺酸在水产动物上的营养生理与免疫作用及其应用已有很多研究，但仍需从以下几个方面开展相关研究：①在"缺乏牛磺酸的植物性蛋白源及某些新型蛋白源广泛应用于水产饲料"的大背景下，进一步评估不同种类水产动物牛磺酸的营养需要情况；②从不同生长阶段及食性等角度，比较研究水产动物牛磺酸合成代谢调控机制；③从与激素的协同作用、改善胰岛素等角度更深入探讨牛磺酸调控水产动物糖代谢的潜在机制；④从绿肝综合征和贫血症等疾病入手深入探究牛磺酸防治水产动物营养性疾病的潜在机理；⑤进一步探讨牛磺酸与其他营养物质的协同或者颉颃作用，如牛磺酸与胱氨酸、半胱氨酸、Zn^{2+}、Cu^{2+} 及主要免疫增强剂等物质之间的相互关系。通过对牛磺酸生理功能及作用机制研究的逐渐深入，牛磺酸在水产动物饲料中的应用潜力将被进一步释放，将为水产养殖业的绿色健康发展奠定坚实基础。

参考文献

[1] Goto T, Tiba K, Sakurada Y, et al. Determination of hepatic cysteine sulfinate decarboxylase activity in fish by means of OPA - prelabeling and reverse - phase high performance liquid chromatographic separation [J]. Fisheries Science, 2001, 67: 553 - 555.

[2] 王蜀金, 陈惠娜, 方思敏, 等. 功能性氨基酸在动物机体内的代谢利用与生理功能 [J]. 家畜生态学报, 2014, 35 (8): 6 - 12.

[3] Spitze A R, Wont D L, Robers Q R, et al. Taurine concentrations in animal feed ingredients: cooking influences taurine content [J]. Journal of Animal Physiology and Animal Nutrition, 2003, 87: 251 - 262.

[4] SalzeE G P, Davis D A. Taurine: a critical nutrient for future fish feeds [J]. Aquaculture, 2015, 437: 215 - 229.

[5] Yokoyama M, Takeuchi T, Park G S, et al. Hepatic cysteine sulphinate decarboxylase activity in fish [J]. Aquaculture Research, 2001, 32 (Suppl. 1): 216 - 220.

[6] Goto T, Matsumoto T, Murakami U S, et al. Conversion of cysteate into taurine in liver of fish [J]. Fisheries Science, 2003, 69: 216 - 218.

[7] Goto T, Matsumoto T, Takagi S. Distribution of the hepatic cysteamine dioxygenase activities in fish [J]. Fisheries Science, 2001, 67: 1187-1189.

[8] 伍琴, 唐建洲, 刘臻, 等. 牛磺酸对鲫鱼 (*Carassius auratus*) 生长、肠道细胞增殖及蛋白消化吸收相关基因表达的影响 [J]. 海洋与湖沼, 2015, 46 (6): 1516-1523.

[9] 刘汝鹏, 樊启学, 全德文, 等. 饲料中豆粕替代鱼粉及添加牛磺酸对翘嘴鲌生长及若干生理指标的影响 [J]. 淡水渔业, 2015, 45 (5): 63-69.

[10] 田芊芊, 胡毅, 毛盼, 等. 低鱼粉饲料中添加牛磺酸对青鱼幼鱼生长、肠道修复及抗急性拥挤胁迫的影响 [J]. 水产学报, 2016, 40 (9): 1330-1339.

[11] 徐璐茜, 明建华, 张易祥, 等. 牛磺酸对草鱼幼鱼生长、非特异性免疫与抗氧化能力以及消化酶活性的影响 [J]. 浙江海洋学院学报 (自然科学版), 2016, 35 (6): 464-471.

[12] 罗莉, 文华, 王琳, 等. 牛磺酸对草鱼生长、品质、消化酶和代谢酶活性的影响 [J]. 动物营养学报, 2006 (3): 166-171.

[13] Li M, Lai H, Li Q, et al. Effects of dietary taurine on growth, immunity and hyperammonemia in juvenile yellow catfish *Pelteobagrus fulvidraco* fed all-plant protein diets [J]. Aquaculture, 2016, 450: 349-355.

[14] 何凌云, 周铭文, 孙云章, 等. 饲料牛磺酸含量对不同生长阶段尼罗罗非鱼生长性能的影响 [J]. 饲料工业, 2018, 39 (22): 14-20.

[15] 王和伟. 饲料牛磺酸水平对吉富罗非鱼和斜带石斑鱼生长的影响 [D]. 厦门: 集美大学, 2013.

[16] 徐奇友, 许红, 郑秋珊, 等. 牛磺酸对虹鳟仔鱼生长、体成分和免疫指标的影响 [J]. 动物营养学报, 2007, (5): 544-548.

[17] Gaylord T G, Barrows F T, Teague A M, et al. Supplementation of taurine and methionine to all-plant protein diets for rainbow trout (*Oncorhynchus mykiss*) [J]. Aquaculture, 2007, 269: 514-524.

[18] 罗志成, 康玉军, 刘哲, 等. 牛磺酸对虹鳟生长性能及生化指标的影响 [J]. 实验室科学, 2020, 23 (6): 37-42.

[19] Wu Y, Ma H, Wang X, et al. Taurine supplementation increases the potential of fishmeal replacement by soybean meal in diets for largemouth bass *Micropterus salmoides* [J]. Aquaculture Nutrition, 2021, 27 (3): 691-699.

[20] 王学习, 周铭文, 黄岩, 等. 饲料牛磺酸水平对不同生长阶段斜带石斑鱼幼鱼生长性能和体成分的影响 [J]. 动物营养学报, 2017, 29 (5): 1810-1820.

[21] Qi G S, Ai Q H, Mai K S, et al. Effects of dietary taurine supplementation to a casein-based diet on growth performance and taurine distribution in two sizes of juvenile turbot (*Scophthalmus maximus* L.) [J]. Aquaculture, 2012, 258-359: 122-128.

[22] Park G S, Takeuchi T, Yokoyama A M, et al. Optimal dietary taurine level for growth of juvenile Japanese flounder *Paralichthys olivaceus* [J]. Fisheries Science, 2002, 68: 824-829.

[23] Kim S K, Matsunari H, Takeuchi T, et al. Comparison of taurine biosynthesis ability between juveniles of Japanese flounder and common carp [J]. Amino Acids, 2008, 35: 161-168.

[24] Martinaz J B, Chatzifotis S, Divanach P, et al. Effect of dietary taurine supplementation on growth performance and feed selection of sea bass *Dicentrarchus labrax* fry fed with demand-feeders [J]. Fisheries Science, 2004, 70: 74-79.

[25] Matsunari H, Yamamoto T, Kim S K, et al. Optimum dietary taurine level in casein-based diet for

juvenile red sea bream *Pagrus major* [J]. Fisheries Science, 2008, 74: 347-353.

[26] Takagi S, Murata H, Goto T, et al. Necessity of dietary taurine supplementation for preventing green liver symptom and improving growth performance in yearling red sea bream *Pagrus major* fed non-fishmeal diets based on soy protein concentrate [J]. Fisheries Science, 2010, 76: 119-130.

[27] 郭斌, 梁萌青, 徐后国, 等. 饲料中添加牛磺酸对红鳍东方鲀幼鱼生长性能、体组成和肝脏中牛磺酸合成关键酶活性的影响 [J]. 动物营养学报, 2018, 30 (11): 4580-4588.

[28] 周婧, 王旭, 刘霞, 等. 饲料中牛磺酸水平对红鳍东方鲀免疫及消化酶的影响 [J]. 大连海洋大学学报, 2019, 34 (1): 101-108.

[29] Lin Y H, Lu R M. Dietary taurine supplementation enhances growth and nutrient digestibility in giant grouper *Epinephelus lanceolatus* fed a diet with soybean meal [J]. Aquaculture Report, 2020, 14: 100464.

[30] Yue R Y, Liu Y J, Tian L X, et al. The effect of dietary taurine supplementation on growth performance, feed utilization and taurine contents in tissues of juvenile white shrimp (*Litopenaeus vannamei* Boone, 1931) fed with low-fishmeal diets [J]. Aquaculture research, 2013, 44: 1317-1325.

[31] 李航, 黄旭雄, 王鑫磊, 等. 饲料中牛磺酸含量对淡水养殖凡纳滨对虾生长、体组成、消化酶活性及抗胁迫能力的影响 [J]. 上海海洋大学学报, 2017, 26 (5): 706-715.

[32] To V A, Liou C H. Taurine supplementation enhances the replacement level of fishmeal by soybean concentrate in diets of juvenile Pacific white shrimp (*Litopenaeus vannamei* Boone, 1931) [J]. Aquaculture Nutrition, 2021, 52 (8): 3771-3781.

[33] 孔圣超, 萧培珍, 朱志强, 等. 牛磺酸对克氏原螯虾生长性能、非特异性免疫以及肝胰脏抗氧化能力的影响 [J]. 中国农学通报, 2020, 36 (11): 136-141.

[34] Sueh B, Gaylin Ⅲ D M. Evaluating the dietary taurine requirement of hybrid striped bass (*Morone chrysops* × *M. saxatilis*) [J]. Aquaculture, 2021, 536: 736473.

[35] 王和伟, 叶继丹, 陈建春. 牛磺酸在鱼类营养中的作用及其在鱼类饲料中的应用 [J]. 动物营养学报, 2013, 25 (7): 1418-1428.

[36] Richard N, Cilen R, Aragao C. Supplementing taurine to plant-based diets improves lipid digestive capacity and amino acid retention of Senegalese sole (*Solea senegalensis*) juveniles [J]. Aquaculture, 2017, 468 (1): 94-101.

[37] Abdeltawwab M, Monier M N. Stimulatory effect of dietary taurine on growth performance, digestive enzymes activity, antioxidant capacity, and tolerance of common carp, *Cyprinus carpio* L. fry to salinity stress [J]. Fish Physiology & Biochemistry, 2018, 44 (2): 1-11.

[38] 王清滨. 牛磺酸对投喂高脂饲料草鱼幼鱼生长、抗氧化能力及脂质代谢的影响 [D]. 长春: 吉林农业大学, 2014.

[39] Lu Y B, You C H, Wang S Q. Physiological changes in Siganus canaliculatus after shallow water stress and the anti-stress effects of taurine [J]. Acta Hydrobiologica Sinica, 2014, 38 (1): 68-74.

[40] Ming J H, Xie J, Xu P, et al. Effects of emodin, vitamin C and their combination on crowding stress resistance of Wuchang bream (*Megalobrama amblycephala*) [J]. Acta Hydrobiologica Sinica, 2011, 35 (3): 400-413.

[41] Zhang Y, Wei Z, Liu G, et al. Synergistic effects of dietary carbohydrate and taurine on growth performance, digestive enzyme activities and glucose metabolism in juvenile turbot *Scophthalmus*

maximus L. [J]. Aquaculture, 2019, 499: 32-41.

[42] 王霜英. 低蛋白饲料中添加牛磺酸对黄河鲤生长性能、体成分及生理生化指标的影响 [D]. 郑州: 河南农业大学, 2015.

[43] 陈晴, 马倩倩, 沈振华, 等. 低鱼粉饲料中补充蛋氨酸、胆汁酸、牛磺酸对中华绒螯蟹幼蟹生长、饲料利用及抗氧化能力的影响 [J]. 海洋渔业, 2018, 40 (1): 65-75.

[44] 邱小琮, 赵红雪. 牛磺酸对鲤生长及血清T3、T4含量的影响 [J]. 淡水渔业, 2006, (1): 22-24.

[45] Kumar P, Prasad Y, Patra A K, et al. Ascorbic acid, garlic extract and taurine alleviate cadmium-induced oxidative stress in freshwater catfish (*Clarias batrachus*) [J]. Science of the Total Environment, 2009, 407: 5024-5030.

[46] 黄岩. 牛磺酸对斜带石斑鱼抗氧化能力的影响 [D]. 厦门: 集美大学, 2017.

[47] Yan L C, Feng L, Jiang D A, et al. Dietary taurine supplementation to a plant protein source-based diet improved the growth and intestinal immune function of young grass carp (*Ctenopharyngodon idella*) [J]. Aquaculture Nutrition, 2019, 25 (4): 1-24.

[48] 吕秋凤, 董公麟, 曹双, 等. 牛磺酸抗应激作用的研究进展 [J]. 中国畜牧杂志, 2014, 50 (21): 78-81.

[49] 石立冬, 翟浩杰, 卫力博, 等. 牛磺酸对红鳍东方鲀热应激转录调控机制的影响 [J]. 中国水产科学, 2020, 27 (10): 1145-1155.

[50] Zhou M, Wu Z H, Liang R S, et al. Effects of dietary taurine, carnitine and cholesterol supplementation on growth performance and immunological status of *Litopenaeus vannamei* under cold exposure [J]. Aquaculture Reasearch, 2017, 48: 1279-1290.

[51] Tan X H, Sun Z Z, Zhu X, et al. Dietary supplementation with taurine improves ability to resist ammonia stress in hybrid snakehead (*Channa maculatus* ♀ × *Channa argus* ♂) [J]. Aquaculture Reasearch, 2018, 49: 3400-3410.

[52] Hu Y, Yang G, Li Z L, et al. Effect of dietary taurine supplementation on growth, digestive enzyme, immunity and resistant to dry stress of rice field eel (*Monopterus albus*) fed low fish meal diets [J]. Aquaculture Research, 2018, 49: 2108-2118.

[53] Gibson G T, Barrows F T, Teague A M, et al. Supplementation of taurine and methionine to all-plant protein diets for rainbow trout (*Oncorhynchus mykiss*) [J]. Aquaculture, 2007, 269: 514-524.

[54] Hosseini S M, HosseiniS A, Eakanda I S, et al. Effect of dietary taurine and methionine supplementation on growth performance, body composition, taurine retention and lipid status of Persian sturgeon, *Acipenser persicus* (Borodin, 1897), fed with plant-based diet [J]. Aquaculture Nutrition, 2018, 24: 324-331.

[55] Couyinbo O F, Simoes R, Monge-Ortiz R, et al. Effects of dietary methionine and taurine supplementation to low-fish meal diets on growth performance and oxidative status of European sea bass (*Dicentrarchus labrax*) juveniles [J]. Aquaculture, 2017, 479: 447-454.

[56] Chen W, Matuda K, Nishimura N, et al. The effect of taurine on cholesterol degradation in mice fed a high-cholesterol diet [J]. Life Sciences, 2004, 74: 1889-1898.

[57] 孟晓雪, 卫育良, 梁萌青, 等. 鱼类胆固醇营养需求研究进展 [J]. 动物营养学报, 2021, 33 (2): 19-728.

[58] Deng J M, Mai K S, Ai Q H, et al. Interactive effects of dietary cholesterol and protein sources on

growth performance and cholesterol metabolism of Japanese flounder (*Paralichthys olivaceus*) [J]. Aquaculture Nutrition, 2010, 16: 419 - 429.

[59] Yun B, Ai Q H, Mai K S, et al. Synergistic effects of dietary cholesterol and taurine on growth performance and cholesterol metabolism in juvenile turbot (*Scophthalmus maximus* L.) fed high plant protein diets [J]. Aquaculture, 2012, 324 - 325: 85 - 91.

[60] Han Y Z, Koshio S, Jiang Z J, et al. Interactive effects of dietary taurine and glutamine on growth performance, blood parameters and oxidative status of Japanese flounder *Paralichthys olivaceus* [J]. Aquaculture, 2014, 434: 348 - 354.

[61] Kals J, Blonk R J W, Vander Mheen H W, et al. Effect of vitamin B_{12} and taurine on the alleviation of nutritional anaemia in common sole (*Solea solea*) [J]. Aquaculture Nutrition, 2019, 25: 456 - 465.

第十四章 谷氨酰胺

谷氨酰胺是水产动物的非必需氨基酸之一，也是水产动物体内含量较为丰富的氨基酸，对维持鱼类肠道健康、保护细胞免受氧化损伤至关重要。本章综述了近年来关于鱼体内谷氨酰胺的合成、代谢和功能的研究进展。这将有助于推进水产动物氨基酸营养的发展，并有利于新型水产饲料的开发与利用。

第一节 谷氨酰胺的理化性质

谷氨酰胺（Glutemine，Gln），化学名2-氨基-4-甲酰胺基丁酸，是谷氨酸的酰胺氨基酸，存在L-谷氨酰胺和D-谷氨酰胺一对对映异构体。谷氨酰胺呈晶性粉末或白色结晶，无臭，稍有甜味。相对分子质量为146.15，熔点为185 ℃，易溶于水，在水中溶解度为42.5 g/L，受热后易分解成谷氨醇或丙酯化为吡咯羧醇。在正常生理pH条件下，谷氨酰胺分子结构中的羧基带负电荷，氨基带正电荷，整个分子对外的净电荷为零。因此，谷氨酰胺也是一种中性氨基酸。

第二节 谷氨酰胺的合成与代谢

一、谷氨酰胺的合成

谷氨酰胺是动物体内含量最丰富的氨基酸之一。谷氨酰胺可以直接从饲料中获得，也可以通过水产动物体内其他氨基酸获得。氨基酸的转氨作用在谷氨酰胺的合成中起着至关重要的作用。在鱼类组织中，谷氨酰胺可由支链氨基酸（α-氨基的供体）和α-酮戊二酸（碳骨架的来源）合成[1]。具体来说，支链氨基酸转氨酶（胞质和线粒体酶）催化支链氨基酸与α-酮戊二酸的转氨化形成谷氨酸，谷氨酰胺合成酶（GS）与氨酯化生成谷氨酰胺。在大多数硬骨鱼的骨骼肌、肝脏和大脑中，谷氨酰胺合成酶仅存在于胞浆中，但该酶在其他组织中的胞内分布尚不清楚，可能也是一种胞浆酶[2]。特别的是，在印度鲇中，谷氨酰胺合成酶主要存在于肝脏线粒体中[3]。在中华乌塘鳢中，谷氨酰胺合成酶在多种细胞类型中表达，其酶活性因组织而异：胃＞肠＞肝＝肌肉[4]。谷氨酰胺虽然能够在机体中自身合成，但在动物机体处于受伤、应激、病理状态以及快速生长条件下时，谷氨酰胺消耗量增加，自身合成的谷氨酰胺不能满足需要，成为机体代谢的限制因素，故而又被称为条件必需性氨基酸。

二、谷氨酰胺的代谢

谷氨酰胺的分解是由一系列酶催化的，包括磷酸盐活化的谷氨酰胺酶、谷氨酸脱氢酶和谷氨酸转氨酶。这些酶的分布在不同的细胞类型、组织和物种之间有很大的差异。谷氨酰胺以谷氨酸和 NH_3 为底物，在谷氨酰胺合成酶的催化下形成，产生 ATP。在鱼类的许多组织，如肝脏、肠道、肾脏和肌肉都具有将 α-酮戊二酸代谢为丙酮酸的酶[5]，随后丙酮酸脱氢酶和 Krebs 循环将其氧化为 CO_2。谷氨酸在鱼类体内的代谢似乎与哺乳动物不同，主要是通过谷氨酸脱氢酶进行脱氨，而大多数谷氨酸在哺乳动物体内被转氨化为天冬氨酸和丙氨酸。在鱼类中，谷氨酰胺和谷氨酸氧化生成 CO_2 的能量转移效率分别为45%和52%，低于脂肪和葡萄糖氧化生成 ATP 的效率（约55%）。在杂交条纹鲈和斑马鱼[1]以及大口黑鲈[6]中，谷氨酸、谷氨酰胺、天冬氨酸和亮氨酸的分解代谢共同促进了肝脏、近端肠、肾脏和骨骼肌中约80%的 ATP 产生。鱼类组织中谷氨酰胺和谷氨酸的氧化速率高于其他营养物质，可能是由于谷氨酸脱氢酶和谷氨酸转氨酶的活性高于葡萄糖和棕榈酸的降解酶。因此，谷氨酰胺和谷氨酸是鱼类的主要能量来源。与哺乳动物一样，鱼类谷氨酰胺水解为谷氨酸以及谷氨酸转化为谷氨酰胺构成了细胞间或细胞内谷氨酰胺-谷氨酸循环[7]。目前，谷氨酰胺或谷氨酸在水生动物体内分解为脯氨酸、瓜氨酸和精氨酸等其他氨基酸的机制尚不清楚[8]。

第三节 谷氨酰胺的转化

研究表明，在某些硬骨鱼中，谷氨酰胺和谷氨酸是瓜氨酸和精氨酸合成的底物[9]，且饲料中的谷氨酰胺可以部分替代鲇饲料中的精氨酸。当鸟氨酸转氨酶（哺乳动物合成精氨酸所需的一种酶）抑制剂后，谷氨酰胺替代部分精氨酸的作用则会消失[10]。然而，杂交条纹鲈饲料中谷氨酰胺不能替代部分精氨酸。谷氨酰胺和脯氨酸是否存在瓜氨酸和精氨酸的物种特异性合成，以及脯氨酸在鱼类组织中的代谢是否存在物种依赖性，还有待进一步研究。

第四节 谷氨酰胺对水产动物生长性能的影响

尽管谷氨酰胺是鱼类组织蛋白和生理体液中含量最为丰富的氨基酸之一，但其在动物体内的营养作用却被长期忽视。近年来，谷氨酰胺在提高鱼类的生长性能、饲料效率和增重率等方面的有益影响已逐渐被报道。以建鲤幼鱼为研究对象，在基础饲料中添加0.40%~2.00%谷氨酰胺，饲喂80 d后，谷氨酰胺添加组显著提高了建鲤幼鱼的增重率、摄食量和饲料效率；但当谷氨酰胺添加量达到1.20%时，继续添加谷氨酰胺对生长性能的促进效果不再增加。在杂交鲟幼鱼中的研究表明，饲料中添加1.00%谷氨酰胺能够显著提高杂交鲟幼鱼末重、特定生长率和饲料效率；而谷氨酰胺添加量为2.00%时，生长性能不再显著提高。Liu 等[11]研究表明，饲料中添加谷氨酰胺显著提高了半滑舌鳎幼虫的存活率和生长性能；以特定生长率为评价指标，半滑舌鳎幼虫的最佳游离谷氨酰胺水平约为 0.63%。Ramos 等[12]研究表明，饲料中添加谷氨酰胺能显著提高巨鲇幼鱼的生长性能，且以 0.48%~0.53%为宜。Macêdo 等[13]研究表明，添加 1.07%谷氨酰胺和谷氨酸可提高罗非鱼生长性能。以大菱鲆为研究对象，在基础饲料中添加 1.50%谷氨酰胺，饲喂 56 d 后，谷氨酰胺添加组显著提高了大菱鲆的末重和饲料效率[14]。此外，谷氨酰胺还可与精氨酸发生协同效应，在组织修复、细胞复制和胶原合成中发挥重要作用，影响动物的生长和生存。研究表明，饲料中添加 2.00%谷氨酰胺和 1.00%精氨酸与谷氨酰胺混合物显著提高了红姑鱼的饲料效率。与基础饲料相比，添加谷氨酰胺和精氨酸组后，红姑鱼中性粒细胞氧化自由基产生量和巨噬细胞的细胞外超氧阴离子均显著提高[15]。目前，谷氨酰胺促进鱼类生长的具体机制仍不明确，在建鲤肠细胞培养实验中发现，在培养液中加入谷氨酰胺能够激活雷帕霉素靶蛋白（TOR）信号通路，提高肠上皮细胞蛋白质合成速率。另外，研究表明饲料中添加谷氨酰胺饲喂 6 周能显著上调建鲤脾脏 TOR 磷酸化水平。TOR 是细胞蛋白质合成调控的关键信号分子，谷氨酰胺促进动物机体生长的功能可能与其对 TOR 信号通路的激活有关。

第五节　谷氨酰胺对水产动物肠道发育与健康的影响

　　尽管谷氨酰胺是一种非必需氨基酸，但在饲料中添加谷氨酰胺可以改善水产动物的肠道健康，因此，谷氨酰胺也被认为是一种功能性氨基酸。目前，已有研究表明谷氨酰胺能够改善肠道结构形态、增强营养物质的跨膜转运，从而提高水产动物的消化吸收能力。当饲料中谷氨酰胺水平为 0.0%~1.0%时，随着谷氨酰胺添加水平的升高，杂交鲟幼鱼肠道和全鱼谷氨酰胺含量和 Na^+/K^+-ATP 酶活性显著升高；但杂交鲟幼鱼肠道蛋白酶、脂肪酶和淀粉酶活性无显著变化[16]。此外，饲料中添加谷氨酰胺也可提高建鲤幼鱼肠道 Na^+/K^+-ATP 酶活性，且谷氨酰胺水平越高，肠道 Na^+/K^+-ATP 酶活性越高；当谷氨酰胺水平超过 1.2%时，各组间消化酶活性无显著差异。此外，饲料中添加谷氨酰胺能够显著提高建鲤肠道重量、肠道皱襞高度，增加蛋白酶、脂肪酶、碱性磷酸酶和 Na^+/K^+-ATP 酶活性[17]。以斑点叉尾鮰幼鱼为研究对象，饲料中谷氨酰胺添加量为 2.0%~3.0%饲喂 10 周后，斑点叉尾鮰肠细胞和微绒毛高度显著提高，并且饲料中添加谷氨酰胺显著提高了肠上皮细胞的迁移率[18]。此外，在哲罗鱼[19]仔鱼和红姑鱼[15]幼鱼中也得到了相似的结论，均表明谷氨酰胺能够改善鱼类肠道形态结构、提高消化酶活力，并对肠道发育具有一定的促进作用。

　　谷氨酰胺作为肠上皮细胞的主要供能物质，对肠细胞的增殖分化起着十分关键的作用。谷氨酰胺对刺激小肠细胞增殖具有积极作用。Jiang 等[20]研究表明，浓度为 2.0~9.2 mmol/L 的谷氨酰胺能促进建鲤肠上皮细胞增殖和分化。Yan 等[21]研究表明，饲料中添加 1.2%谷氨酰胺可促进建鲤幼鱼肠道发育和功能，预防肠道上皮损伤。其原因可能与肠道细胞分裂的差异有关，也可能与样品的处理、鱼类种类、年龄等有关。谷氨酰胺对消化酶活性的积极作用可能与饲料中谷氨酰胺水平的提高促进了鲤肝胰脏的发育有关。除鲟外，大多数鱼类没有隐窝，肠道酶主要来自肝胰脏。肝胰脏的酶分泌与肝胰脏的发育密切相关。饲料中谷氨酰胺含量越高，肝胰脏指数和蛋白质含量越高，且添加谷氨酰胺后酶活性的提高与鲤肝胰脏的发育呈正相关[21]。添加谷氨酰胺显著降低了固有层和黏膜下层白细胞的浸润和炎症细胞因子 IL-8、TNF-α 和 TGF-β 的表达水平，提高乳酸菌和芽孢杆菌的相对丰度，激活先天免疫系统和适应性免疫系统，改变肠道黏膜微生物群落结构[21]。

谷氨酰胺还是一种有效的肠道保护剂，可以通过抑制炎症反应来缓解豆粕性肠炎等负面影响。将原代培养的大菱鲆肠上皮细胞与 1 mg/mL 的大豆皂苷和不同水平的谷氨酰胺（0、0.5、1.0 和 2.0 mmol/L）同时给药 12 h 发现，谷氨酰胺显著降低了大菱鲆肠上皮细胞中 IL-1β、IL-8 和 TNF-α 的表达，显著降低了细胞核和胞浆中 NF-κB p65 丰度，并呈剂量依赖关系，升高了 IκBα 蛋白水平，但降低了其磷酸化水平。下调 IKKα/β 和磷酸化 IKKα/β 水平，减弱了大豆皂苷诱导的大比目鱼肠道的炎性反应[22]。此外，饲料中添加谷氨酰胺还可通过不同途径抑制大豆皂苷引起的氨代谢和肠道损伤：①饲料中谷氨酰胺可能通过 NF-κB、AP-1 和 MAPKs 信号通路，通过降低白细胞的含量和抑制炎症细胞因子的过表达，对大豆皂苷诱导的肠道炎症起保护作用。②饲料中谷氨酰胺通过 JNK 和 p38 通路调控紧密连接蛋白。③谷氨酰胺通过介导大豆皂苷诱导的氧化应激和抗氧化酶基因及抗氧化相关信号因子 Nrf2 的表达来缓解氧化应激。④谷氨酰胺可能具有将自噬过程控制在适当水平上的作用[23]。He 等[24]研究发现，饲料中添加谷氨酰胺能够缓解由大豆皂苷诱导的杂交石斑鱼的肠道代谢。谷氨酰胺促进了肠道发育、细胞增殖和肠道免疫，并降低了炎性细胞因子原蛋白 TNF-α、IL-1β 和 IFN-α 的基因表达。这可能是由于饲料中添加谷氨酰胺可抑制 MyD88/NF-κB 信号通路，引起该通路导致炎性细胞因子的分泌，并改变肠道菌群结构，最终促进肠道生长。

第六节　谷氨酰胺对水产动物抗氧化能力的影响

氧化损伤导致消化器官发育不良，营养物质消化和吸收受损，生长受限。体外研究结果表明，2~6 mmol/L 谷氨酰胺可提高鲤肠上皮细胞的抗氧化能力，并调节过氧化氢酶、谷胱甘肽还原酶和谷胱甘肽过氧化物酶等抗氧化酶的基因表达水平[25]，并保护其肠上皮细胞免受铜诱导的氧化损伤[26]。饲料中添加谷氨酰胺还能降低草鱼肠道脂肪过氧化和氧化损伤[27]。谷氨酰胺结合其他氨基酸保护鲤红细胞免受氧化损伤，这可能是保护红细胞免受凋亡的主要因素[28]。研究表明，谷氨酰胺对 H_2O_2 诱导的鲤肠上皮细胞氧化应激有一定的保护作用[29]。此外，Hu 等[30]研究证实，谷氨酰胺可修复受损的脂质和蛋白质，修复了 H_2O_2 诱导的鱼肠上皮细胞损伤，维持了细胞的增殖、分化和功能，这可能与恢复的谷胱甘肽含量和抗氧化酶活性有关。

Zhu 等[31]研究表明，1.2% 和 1.5% 谷氨酰胺添加组血清、肝胰腺和肌肉中谷胱甘肽过氧化物酶（GPx）、超氧化物歧化酶（SOD）活性和谷胱甘肽（GSH）含量显著升高，血清丙二醛（MDA）水平显著降低。由此可见，饲料中添加 0.9%~1.2% 的谷氨酰胺可增强杂交鲟幼鱼的抗氧化防御系统活性和血清非特异性免疫功能。当饲料谷氨酰胺水平为 1.0% 时，杂交鲟幼鱼肠道或全鱼中超氧化物歧化酶活性显著升高，丙二醛水平显著降低[32]。摄食高水平谷氨酰胺后，牙鲆幼鱼表现出更高的抗氧化能力，这说明谷氨酰胺在鱼类氧化应激保护中发挥着重要作用[33]。Jiang 等[25]研究表明，谷氨酰胺可逆转饲料中暴露的甘氨酸（80 g/kg）引起的建鲤肠道氧化损伤和肠道物理屏障和功能破坏。Chen 等[29]证实，谷氨酰胺能完全阻断 H_2O_2 刺激下乳酸脱氢酶的释放，还可降低由 H_2O_2 诱导的丙二醛水平和蛋白质羰基含量的增加。谷氨酰胺完全抑制了 H_2O_2 诱导的 Na^+/K^+-ATP 酶、超氧化物歧化酶、过氧化氢酶、谷胱甘肽过氧化物酶、谷胱甘肽还原酶、谷胱甘肽转移酶活性的下降，降低了还原性谷胱甘肽浓度和还原性谷胱甘肽与氧化性谷胱甘肽比值。Coutinho 等[34]评价了饲料中添加谷氨酰胺对金头鲷肝脏和肠道谷氨酰胺代谢及氧化状态的影响。饲料中添加 0.5%～2.0%谷氨酰胺饲喂 6 周后，金头鲷肝脏表现出较高的过氧化氢酶和葡萄糖-6-磷酸脱氢酶活性，而肠道表现出较高的谷胱甘肽过氧化物酶和谷胱甘肽还原酶活性和氧化型谷胱甘肽含量，这表明肠道抗氧化反应具有较高的谷胱甘肽依赖性。饲料中添加谷氨酰胺可提高肝脏和肠道总谷胱甘肽和还原性谷胱甘肽含量以及肠道超氧化物歧化酶活性，但对脂质过氧化值无影响，阐明了饲料中添加谷氨酰胺对金头鲷抗氧化反应的潜力。

第七节　谷氨酰胺对水产动物的其他影响

目前，谷氨酰胺对雄鱼或雌鱼生育能力的影响尚不清楚。然而，体外研究的结果显示，2 mmol/L 的谷氨酰胺可以提升鱼精原细胞的增殖、成熟和功能（如受精卵）[35]。也有证据表明，添加 5.1 mmol/L 谷氨酰胺可提高虹鳟精子的活力和虹鳟卵的受精率；添加 1 μmol/L 谷氨酸可增加虹鳟卵巢卵泡的甾体生成[36]。此外，谷氨酰胺及其代谢物氨基丁酸还能刺激鱼类释放促黄体生成素[37]，这可能与促性腺激素释放激素的释放有关。

第八节　小结与展望

在过去的几十年中，人们对鱼类中谷氨酰胺的营养和代谢越来越感兴趣。随着水产养殖在为人类提供高质量蛋白质方面发挥着越来越重要的作用，谷氨酰胺对推动氨基酸营养领域的发展具有及时和重要的意义。饲料中的大部分氨基酸以蛋白结合形式存在。在植物蛋白饲料中添加谷氨酰胺能有效改善鱼的生长、代谢和免疫力。这意味着人们可以通过谷氨酰胺营养策略减少水产饲料中鱼粉的使用。此外，由于饲料蛋白质通常含有 16%的氮，使用谷氨酰胺有望减少鱼类的氮排泄[38]。尽管许多研究表明谷氨酰胺对鱼类生长和发育有一些积极影响，但鱼类对谷氨酰胺的需要量尚未确定，这一问题应在今后的研究中加以解决。此外，结晶氨基酸具有较高的从饲料颗粒浸出到周围水体的潜力，可能与蛋白质消化释放的氨基酸吸收不同步。以大菱鲆为对象的研究结果表明，晶体氨基酸可以替代饲料中高达 19%的蛋白质，而不会对大菱鲆的生长性能或饲料利用效率产生负面影响，但游离氨基酸替代较高水平的饲料蛋白质严重抑制了大菱鲆的生长性能[39]。目前尚不清楚这种吸收不同步是否会对其他水生动物的生长、饲料利用和生理状况产生负面影响。为了最大限度地减少饲料中游离谷氨酰胺的损失，一种可行的方法可能是用脂质包裹它们。未来的研究需要评估不同形式的氨基酸（包括游离氨基酸与结晶氨基酸）应用于不同物种的效率。

参考文献

[1] Jia S, Li X, Zheng S, et al. Amino acids are major energy substrates for tissues of hybrid striped bass and zebrafish [J]. Amino Acids, 2017, 49 (12): 2053-2063.

[2] Wright P A, Steele S L, Huitema A, et al. Induction of four glutamine synthetase genes in brain of rainbow trout in response to elevated environmental ammonia [J]. Journal of Experimental Biology, 2007, 210 (16): 2905-2911.

[3] Saha N, Das L, Dutta S. Types of carbamyl phosphate synthetases and subcellular localization of urea cycle and related enzymes in air-breathing walking catfish, *Clarias batrachus* [J]. Journal of Experimental Zoology, 1999, 283 (2): 121-130.

[4] Anderson P M, Broderius M A, Fong K C, et al. Glutamine synthetase expression in liver, muscle, stomach and intestine of *Bostrichthys sinensis* in response to exposure to a high exogenous ammonia concentration [J]. Journal of Experimental Biology, 2002, 205 (14): 2053-2065.

[5] Chamberlin M E, Glemet H C, Ballantyne J S. Glutamine metabolism in a holostean (*Amia calva*) and teleost fish (*Salvelinus namaycush*) [J]. American Journal of Physiology, 1991, 260 (1-2): R159-R166.

[6] Li X, Shixuan Z, Jia S, et al. Oxidation of energy substrates in tissues of largemouth bass (*Micropterus salmoides*) [J]. Amino Acids, 2020, 52 (6-7): 1017-1032.

[7] Li P, Wu G. Roles of dietary glycine, proline, and hydroxyproline in collagen synthesis and animal growth [J]. Amino Acids, 2018, 50 (1): 29-38.

[8] Li P, Mai K, Trushenski J, et al. New developments in fish amino acid nutrition: towards functional and environmentally oriented aquafeeds [J]. Amino Acids, 2009, 37 (1): 43-53.

[9] Buentello J A, Gatlin D M. The dietary arginine requirement of channel catfish (*Ictalurus punctatus*) is influenced by endogenous synthesis of arginine from glutamic acid [J]. Aquaculture, 2000, 188 (3): 311-321.

[10] Buentello J A, Gatlin D M. Plasma citrulline and arginine kinetics in juvenile channel catfish, *Ictalurus punctatus*, given oral gabaculine [J]. Fish Physiology and Biochemistry, 2001, 24 (2): 105-112.

[11] Liu J, Mai K, Xu W, et al. Effects of dietary glutamine on survival, growth performance, activities of digestive enzyme, antioxidant status and hypoxia stress resistance of half-smooth tongue sole (*Cynoglossus semilaevis* Günther) post larvae [J]. Aquaculture, 2015, 446: 48-56.

[12] Palomino Ramos A R, Campelo D A V, Carneiro C L D S, et al. Optimal dietary L-glutamine level improves growth performance and intestinal histomorphometry of juvenile giant trahira (*Hoplias lacerdae*), a Neotropical carnivorous fish species [J]. Aquaculture, 2022, 547: 737469.

[13] Macêdo É S D, Franco T S G, Natali M R M, et al. Dietary glutamine-glutamate supplementation enhances growth performance and intestinal villi development in cage-farmed Nile tilapia fingerlings [J]. Revista Brasileira de Zootecnia, 2021, 50.

[14] Gu M, Bai N, Xu B, et al. Protective effect of glutamine and arginine against soybean meal-induced enteritis in the juvenile turbot (*Scophthalmus maximus*) [J]. Fish & Shellfish Immunology, 2017, 70: 95-105.

[15] Cheng Z, Buentello A, Gatlin D M. Effects of dietary arginine and glutamine on growth performance, immune responses and intestinal structure of red drum, *Sciaenops ocellatus* [J]. Aquaculture, 2011, 319 (1): 247-252.

[16] 王常安, 李晋南, 王连生, 等. L-丙氨酰-L-谷氨酰胺对杂交鲟体成分、肠道形态的影响 [J]. 水产学杂志, 2017, 30 (2): 12-16.

[17] Hu K, Zhang J X, Feng L, et al. Effect of dietary glutamine on growth performance, non-specific immunity, expression of cytokine genes, phosphorylation of target of rapamycin (TOR), and anti-oxidative system in spleen and head kidney of Jian carp (*Cyprinus carpio* var. Jian) [J]. Fish Physiology and Biochemistry, 2015, 41 (3): 635-649.

[18] Pohlenz C, Buentello A, Mwangi W, et al. Arginine and glutamine supplementation to culture media improves the performance of various channel catfish immune cells [J]. Fish & Shellfish Immunology, 2012, 32 (5): 762-768.

[19] 徐奇友, 王常安, 许红, 等. 丙氨酰-谷氨酰胺对哲罗鱼仔鱼生长和抗氧化能力的影响 [J]. 动物营养学报, 2009, 21 (6): 1012-1017.

[20] Jiang J, Zheng T, Zhou X Q, et al. Influence of glutamine and vitamin E on growth and antioxidant capacity of fish enterocytes [J]. Aquaculture Nutrition, 2009, 15 (4): 409-414.

[21] Yan L, Zhou X Q. Dietary glutamine supplementation improves structure and function of intestine of juvenile Jian carp (*Cyprinus carpio* var. Jian) [J]. Aquaculture, 2006, 256 (1): 389-394.

[22] Gu M, Pan S, Deng W, et al. Effects of glutamine on the IKK/Ikappa B/NF-small ka, Cyrillic B system in the enterocytes of turbot *Scophthalmus maximus* L. stimulated with soya-saponins [J]. Fish & Shellfish Immunology, 2021, 119: 373-378.

[23] Gu M, Pan S, Li Q, et al. Protective effects of glutamine against soy saponins-induced enteritis, tight junction disruption, oxidative damage and autophagy in the intestine of *Scophthalmus maximus* L. [J]. Fish & Shellfish Immunology, 2021, 114: 49-57.

[24] He Y, Liang J, Dong X, et al. Glutamine alleviates β-conglycinin-induced enteritis in juvenile hybrid groupers *Epinephelus fuscoguttatus* ♀ × *Epinephelus lanceolatus* ♂ by suppressing the MyD88/NF-κB pathway [J]. Aquaculture, 2022, 549: 737735.

[25] Jiang W D, Hu K, Zhang J X, et al. Soyabean glycinin depresses intestinal growth and function in juvenile Jian carp (*Cyprinus carpio* var Jian): protective effects of glutamine [J]. British Journal of Nutrition, 2015, 114 (10): 1569-1583.

[26] Jiang J, Wu X Y, Zhou X Q, et al. Glutamate ameliorates copper-induced oxidative injury by regulating antioxidant defences in fish intestine [J]. British Journal of Nutrition, 2016, 116 (1): 70-79.

[27] Zhao Y, Hu Y, Zhou X Q, et al. Effects of dietary glutamate supplementation on growth performance, digestive enzyme activities and antioxidant capacity in intestine of grass carp (*Ctenopharyngodon idella*) [J]. Aquaculture Nutrition, 2015, 21 (6): 935-941.

[28] Li H T, Feng L, Jiang W D, et al. Oxidative stress parameters and anti-apoptotic response to hydroxyl radicals in fish erythrocytes: protective effects of glutamine, alanine, citrulline and proline [J]. Aquatic Toxicology, 2013, 126: 169-179.

[29] Chen J, Zhou X, Feng L, et al. Effects of glutamine on hydrogen peroxide-induced oxidative damage in intestinal epithelial cells of Jian carp (*Cyprinus carpio* var. Jian) [J]. Aquaculture, 2009, 288 (3): 285-289.

[30] Hu K, Feng L, Jiang W, et al. Oxidative damage repair by glutamine in fish enterocytes [J]. Fish Physiology and Biochemistry, 2014, 40 (5): 1437–1445.

[31] Zhu Q, Xu Q Y, Xu H, et al. Dietary glutamine supplementation improves tissue antioxidant status and serum non–specific immunity of juvenile Hybrid sturgeon (*Acipenser schrenckii* ♀ × *Huso dauricus* ♂) [J]. Journal of Applied Ichthyology, 2011, 27 (2): 715–720.

[32] Wang C A, Xu Q Y, Xu H, et al. Dietary L–alanyl–L–glutamine supplementation improves growth performance and physiological function of hybrid sturgeon *Acipenser schrenckii* ♀ × *A. baerii* ♂ [J]. Journal of Applied Ichthyology, 2011, 27 (2): 727–732.

[33] Han Y, Koshio S, Jiang Z, et al. Interactive effects of dietary taurine and glutamine on growth performance, blood parameters and oxidative status of Japanese flounder *Paralichthys olivaceus* [J]. Aquaculture, 2014, 434: 348–354.

[34] Coutinho F, Castro C, Rufino–Palomares E, et al. Dietary glutamine supplementation effects on amino acid metabolism, intestinal nutrient absorption capacity and antioxidant response of gilthead sea bream (*Sparus aurata*) juveniles [J]. Comparative Biochemistry and Physiology. Part A, Molecular and Integrative Physiology, 2016, 191: 9–17.

[35] Higaki S, Shimada M, Kawamoto K, et al. *In vitro* differentiation of fertile sperm from cryopreserved spermatogonia of the endangered endemic cyprinid honmoroko (*Gnathopogon caerulescens*) [J]. Scientific Reports, 2017, 7: 42852.

[36] Leatherland J F, Lin L, Renaud R. Effect of glutamate on basal steroidogenesis by ovarian follicles of rainbow trout (*Oncorhynchus mykiss*) [J]. Comparative Biochemistry and Physiology. B: Biochemistry and Molecular Biology, 2004, 138 (1): 71–80.

[37] Trudeau V L, Spanswick D, Fraser E J, et al. The role of amino acid neurotransmitters in the regulation of pituitary gonadotropin release in fish [J]. Biochemistry and Cell Biology, 2000, 78 (3): 241–259.

[38] Sano C. History of glutamate production [J]. American Journal of Clinical Nutrition, 2009, 90 (3): 728S–732S.

[39] Peres H, Oliva–Teles A. The effect of dietary protein replacement by crystalline amino acid on growth and nitrogen utilization of turbot *Scophthalmus maximus* juveniles [J]. Aquaculture, 2005, 250 (3): 755–764.

第十五章 多 肽

肽（Peptide）是指分子结构介于氨基酸和蛋白质之间的一类化合物，是由 20 种天然氨基酸以不同组成和排列方式构成的，从二肽到复杂的线性或环形结构的多肽的总称。每一种肽都具有独特的组成结构，不同肽的组成结构决定了其功能。肽在生物体内的含量是很微量的，但却具有显著的生理活性。其中可调节生物体生理功能的肽称为功能肽（functional peptide）或生物活性肽（biologically active 或 biopeptide）。作为机体先天免疫系统的重要组成部分，生物活性肽具有抵抗细菌、真菌、病毒和寄生虫等多种微生物病原体的能力。在大多数情况下，生物活性肽对病原体的特异性较低，不易产生耐药性。此外，部分生物活性肽还具有中和内毒素、促进血管生成、促进伤口愈合等调节作用[1]。鱼类是生物活性肽的重要来源，目前已经在鱼类中发现了 150 余种生物活性肽。已经被分离出来的鱼源生物活性肽除了对其抗菌功能进行研究外，只有一小部分对其潜在应用进行了研究。因此，本章重点阐明鱼源生物活性肽的分类、生物学活性、作用机制及其潜在的应用前景，这不仅能为鱼源生物活性肽的产业化应用提供理论参考，也对水产养殖业的饲料开发与资源利用具有重要意义。

第一节 鱼源生物活性肽的分类

一、具有 α-螺旋结构的鱼源生物活性肽

具有 α-螺旋结构的鱼源生物活性肽一般含有 13~50 个氨基酸残基，主要包含 piscidins、pardaxins 和其他 α-螺旋鱼源生物活性肽。α-螺旋鱼源生物活性肽的物种来源、氨基酸序列和活性特征等信息见表 15-1。

1. Piscidins　Piscidins 是一类具有典型 α-螺旋结构的阳离子生物活性肽，一般由 22~46 个氨基酸残基组成，且序列中含有大量的苯丙氨酸（F）、组氨酸（H）和异亮氨酸（I）[2]。Piscidins 在软骨细胞、肝脏、皮肤和鳃等多种器官中均有表达，但不同亚型的生物活性肽表达量往往也不尽相同[3]。目前发现的 piscidins 主要有三类：Ⅰ类 piscidins 具有 22 个氨基酸残基，即 piscidin-1、piscidin-2 和 piscidin-3，对细菌和纤毛原生动物具有广泛的活性，仅在肥大细胞中被分离；Ⅱ类 piscidins 通常含有 44~46 个氨基酸残基，对细菌和原生动物具有活性，但其活性水平处于Ⅰ和Ⅲ类之间；Ⅲ类 piscidins 含有 55 个氨基酸残基，对原生动物具有较高的活性[4]。Piscidins 的抗菌活性主要依赖于氨基酸序列中含有的组氨酸[5]。此外，部分 piscidins 还具有抗真菌及抗肿瘤的能力[2]。

表 15-1 具有 α-螺旋结构的鱼生物活性肽

家族	物种	序列	活性	参考文献
Piscidins				
Piscidin-1	*Morone saxatilis*	FFHHIFRGIVHVGKTIHRLVTG	细菌 G^+/G^-、真菌	Lauth 等[19]
Piscidin-2	*Morone saxatilis*	FFHHIFRGIVHVGKTIHKLVTG	细菌 G^+/G^-、真菌、病毒	Gampagna 等[20]
Piscidin-3	*Morone chrysops* × *Morone saxatilis*	FIHHIFRGIVHAGRSIGRFLTG	细菌 G^+/G^-	Silphaduang 等[21]
Piscidin-4	*Morone chrysops* × *Morone saxatilis*	FFRHLFRGAKAIFRGARQGXRAHKVVSRYRNRDVPETDNNQEEP	细菌 G^+/G^-、寄生虫	Noga 等[22]
Piscidin-5	*Morone chrysops* × *Morone saxatilis*	LIGSLFRGAKAIFRGARQGWRSHKAVSRYRARYVRRPVIYYHRVYP	细菌 G^+/G^-、寄生虫	Salger 等[23]
Piscidin-6	*Morone chrysops* × *Morone saxatilis*	LFGSVKAWFKGGKKGFQDYRYQKDMAKMNKRYGPNWQQRGGQEPPADAQANDQPP	细菌 G^+/G^-、寄生虫	Salger 等[23]
Piscidin-7	*Morone chrysops* × *Morone saxatilis*	FFGRLKSMWRGARGGLKAYKYWKDMAKMNKRYGPNWQQGGGQEPPADAQANDQPP	细菌 G^+/G^-、寄生虫	Salger 等[23]
Moronecidins				
Moronecidin	*Morone chrysops*	FFHHIFRGIVHVGKTIHKLVTG	细菌 G^+/G^-、真菌	Lauth 等[19]
Moronecidin	*Morone saxatilis*	FFHHIFRGIVHVGKTIHRLVTG	细菌 G^+/G^-、真菌	Lauth 等[19]
Dicentracin	*Dicentrarchus labrax*	FFHHIFRGIVHVGKSIHKLVTG	—	Salerno 等[24]
Pleurocidins				
WF1/2	*P. americanus*	GWGSFFKKAAHVGKHVGKAALTHYL（G）	细菌 G^+/G^-	Cole 等[25]
WF1L	*P. americanus*	GKGRWLDRIGKAGGIIIGGALDHL	细菌 G^+/G^-、真菌	Douglas 等[26]
WFX	*P. americanus*	RSTEDIIKSISGGGFLNAMNA	细菌 G^+/G^-、真菌	Douglas 等[26]
WFY	*P. americanus*	FLGFLFHGIRHGIKAIHGMIH	细菌 G^+/G^-、真菌	Douglas 等[26]
WF3	*P. americanus*	FLGALIKGAIHGGRFIHGMIQNH	细菌 G^+/G^-、真菌	Douglas 等[27]
WF4	*P. americanus*	GWGSIFKHGRHAAKHIGHAAVNHYL	细菌 G^+/G^-、真菌	Douglas 等[27]
Pardaxins				
Pardaxin	*Pardachirus marmoratus*	GFFALIPKIISSPLFKTLLSAVGSALSSSGGQE	细菌 G^+/G^-、真菌	Pan 等[9]

(续)

家族	物种	序列	活性	参考文献
Gaduscidins				
Gaduscidin-1	*Gadus morhua*	FIHHIIGWISHCVRAIHRAIH	细菌 G^+/G^-	Browne 等[13]
Gaduscidin-2	*Gadus morhua*	FLHHIVGLIHHGLSLFGDR	细菌 G^+/G^-	Browne 等[13]
Grammistins				
Grammistin Pp 2b/Gs E	*Pogonoperca punctata*	FIGGIISFIKKLF	细菌 G^+/G^-	Sugiyama 等[28]
Grammistin Pp 2a	*Pogonoperca punctata*	FIGGIISLIKKLF	细菌 G^+/G^-	Kaji 等[29]
Grammistin Pp 4b	*Pogonoperca punctata*	LFGFLIPLLPHLIGAIPQVIGAIR	细菌 G^+/G^-	Kaji 等[29]
Grammistin Pp 4a/Gs G	*Pogonoperca punctata*	LFGFLIPLLPHIIGAIPQVIGAIR	细菌 G^+/G^-	Kaji 等[29]
Grammistin Pp 3	*Pogonoperca punctata*	NWRKILGQIASVGAGLLGSLLAGYE	细菌 G^+/G^-	Sugiyama 等[28]
Grammistin Pp 1/Gs D	*Pogonoperca punctata*	FIGGIISFFKRLF	细菌 G^+/G^-	Sugiyama 等[28]
Grammistin Gs F	*Grammistes sexlineatus*	LFGFLIKLIPSLFGALSNIGRNRNQ	细菌 G^+/G^-	Sugiyama 等[28]
Grammistin Gs C	*Grammistes sexlineatus*	NWRKILGKIAKVAAGLLCSMLAGYQV	细菌 G^+/G^-	Sugiyama 等[28]
Grammistin Gs A	*Grammistes sexlineatus*	WWRELLKKLAFTAAGHLGSVLAAKQSGW	细菌 G^+/G^-	Sugiyama 等[28]
Grammistin Gs B	*Grammistes sexlineatus*	IGGIISFFKRLF	细菌 G^+/G^-	Sugiyama 等[28]
Epinecidin				
Epinecidin	*Epinephelus coioides*	FIFHIIKGLFHAGKMIHGLVTRRRH	细菌 G^+/G^-	Pan 等[15]
Chrysophsins				
Chrysophsin-1	*Pagrus major*	FFGWLIKGAIHAGKAIHGLIHRRRH	细菌 G^+/G^-	Mason 等[17]
Chrysophsin-2	*Pagrus major*	FFGWLIRGAIHAGKAIHGLIHRRRH	细菌 G^+/G^-	Saitoh 等[30]
Chrysophsin-3	*Pagrus major*	FIGLLISAGKAIHDLIRRRH-	细菌 G^+/G^-	Saitoh 等[30]

注：G^+ 代表革兰氏阴性菌，G^- 表示革兰氏阴性菌。下表同。

Moronecidins、pleurocidins 和 piscidins 之间具有相似的基因结构，同属于一类肽[6]。Moronecidins 和 piscidins 几乎是同时在条纹鲈中被发现。Pleurocidins 处于 N-末端区域的氨基酸残基往往比 C-末端区域的氨基酸残基对 pleurocidins 的抗菌活性更为重要；其中，位于 N-端区域的 Trp-2、Phe-5 和 Phe-6 残基在肽插入细胞膜中起着主要作用[7]。

2. Pardaxins Pardaxins 最初是从豹鳎鱼背鳍及臀鳍腺体中分离的一类离子型神经毒素，其序列中一般含有 33 个氨基酸残基[8]。目前发现的 pardaxins 主要有五种，各种类之间具有高度的序列同源性，其氨基酸残基仅在第 5、14 或 31 号位上存在不同[6]。除抗细菌活性外，pardaxins 还可抑制各种癌细胞，包括人纤维肉瘤[9]、乳腺癌[10]和犬肛周腺瘤[11]等。此外，pardaxins 对绵羊红细胞表现出一定的溶血活性，且不同种类的 pardaxins 溶血性活性也不尽相同，如 pardaxin-1 的细胞毒性是 pardaxin-2 的 5~10 倍[12]。

3. 其他 α-螺旋鱼源生物活性肽 人们还发现了几种与 piscidins 和 pardaxins 序列同源性较远的生物活性肽，分别是 gaduscidins、grammistins、epinecidin 和 chrysophsins。

Gaduscidins 是一类富含组氨酸的生物活性肽，最早在大西洋鳕中被分离，在头肾、血液、脑、鳃、幽门盲肠和脾脏中均可表达[13]。目前分离出的 gaduscidins 主要有两种：GAD-1 和 GAD-2。GAD-1 对细菌具有广泛的活性；而 GAD-2 对细菌则无活性，但 GAD-2 表现出更高的抗寄生虫活性[14]。Gaduscidins 与 moronecidins 的同源关系较远，这可能是其序列中缺少谷氨酰胺（Q）组成的基序导致的[13]。

Grammistins 具有高度的两亲性，对革兰氏阳性菌和革兰氏阴性菌均具有良好的抗菌活性。根据序列同源性，grammistins 可分为三个亚类：Ⅰ类通常含有 24~25 个氨基酸残基，即 Gs 1、Pp 4a（Gs 2）和 Pp 4b；Ⅱ类含有 12~13 个氨基酸残基，即 Gs B、Pp 1（Gs D）、Pp 2a、Pp 2b（Gs E）；Ⅲ类含有 25~28 个氨基酸残基，即 Gs A、Gs C 和 Pp 3。除 Gs A、Gs B 和 Gs C 外，其余 grammistins 均具有不同程度的溶血活性[6]。

Epinecidin 是从斜带石斑鱼的鳃和肠道中分离到的一类能够保护鱼类免受病原菌感染的生物活性肽，通常含有 21 个氨基酸残基[15]。Epinecidin 除对大多数水产病原菌具有很强的抑菌活性外，还能够促进伤口愈合、血管新生，抑制血液中细菌数量，降低免疫相关炎症因子的基因表达水平[16]。

Chrysophins 最初是从真鲷鳃中分离纯化得到，肽链长度一般为 19~25 个氨基酸残基[17]。目前已知的 chrysophsins 主要有三种：chrysophsin-1、chrysophsin-2 和 chrysophsin-3。C-末端部分由精氨酸（R）和组氨酸组成的 RRRH 氨基酸基序可以干扰细胞膜的磷脂双分子层，在肽插入脂质膜的过程中起着关键作用[18]。

二、含有二硫键的鱼源生物活性肽

鱼源生物活性肽按照含有二硫键的数量不同可分为 cathelicidins、防御素（defensins）和 hepcidins。含有二硫键的鱼源生物活性肽的物种来源、氨基酸序列和活性特征等信息见表 15-2。

表 15-2 含有二硫键结构的鱼源生物活性肽

家族	物种	序列	活性	参考文献
Cathelicidins				
HFIAP-1	*Myxine glutinosa*	GFFKKAWRKVKHAGRRVLDTAKGVGRHYVNNWLNRYRZ	细菌 G⁺/G⁻、真菌	Uzzell 等[54]
HFIAP-3	*Myxine glutinosa*	GWFKKAWRKVKNAGRRVLKGVGIHYGVGLI	细菌 G⁺/G⁻	Uzzell 等[54]
CATH$_{BR}$ALE	*Brachymystax lenok*	RRSKARGGSRGSKMGRKDSKGGSRGRPGSGSRPGGSSIAGASRGDRGGTRNA	细菌 G⁺/G⁻	厉政策[34]
CodCath	*Gadus morhua*	SRSGRGSGKGGRGSRGSSGSRGSKGPSGSRGSSGSRGSKGSRGRGSTIAGNRNNGCTRTA	细菌 G⁺	Broekman 等[55]
aCath	*Plecoglossus altivelis*	RMRRSKSGKGSGSKGSGSKGSGSKGSGSKGSRPGGSSIAGGSKGKGTQTA	细菌 G⁻	Lu 等[56]
防御素				
Onβ-defensin*	*Oreochromis niloticus*	MFCYRVVLALLVLLLNVAQNEAASFPWSCLSLSGVCRRKVCLPTELFFGPLCCGKGSLCCVSHFL	细菌 G⁺/G⁻	Dong 等[57]
Scβ-defensin	*Siniperca chuatsi*	MKGLSLVLLVLLLMLAVGEGNDPEMQYWTCGYRGLCRRFCYAQEYIVGHHCPRRYRCCAMRS	细菌 G⁺/G⁻	Wang 等[58]
zfβ-defensin1	*Danio rerio*	MKPQSFILLVLVVLALHFKENEA	细菌 G⁺/G⁻	Zou 等[59]
zfβ-defensin2	*Danio rerio*	MKKLGMIIFTLLALFAGNVHN	细菌 G⁺/G⁻	Zou 等[59]
tnβ-defensin2	*Tetraodon nigroviridis*	MKGKSLVLLVLLLMLAAG	细菌 G⁺/G⁻	Zou 等[59]
fuβ-defensin1	*Takifugu rubripes*	MASYRAVVLALLVLVLNAVENEA	细菌 G⁺/G⁻	Zou 等[59]
ecDefensin	*Epinephelus coioides*	NDPEMQYWTCGYRGLCRRFCHAQEYIVGHHCPRRYRCCAVRS	细菌 G⁺/G⁻	Guo 等[60]
Hepcidins				
Hepcidin	*Sparus aurata*	CRFCCRCCPRMRGCGLCCRF	细菌 G⁺/G⁻	Cuesta 等[61]
Hepcidin*	*Scophthalmus maximus*	MKCKPCCNCCNLNGCGVCCDF	细菌 G⁺/G⁻	Pereiro 等[62]
Hepcidin	*Danio rerio*	LCRFCCKCCRNKGCGYCCKF	细菌 G⁺/G⁻	Shike 等[63]
Hepcidin	*Morone saxatilis* × *M. chrysops*	GCRFCCNCCPNMSGCGVCCRF	细菌 G⁺/G⁻	Lauth 等[64]
AS-hepc6*	*Acanthopagrus schlegelii*	CRFCCRCCPRMRGCGLCCRF	细菌 G⁺/G⁻	Cuesta 等[61]

注：*表示该肽氨基酸序列由基因序列推测得出。

1. Cathelicidins　Cathelicidins 是一类具有 cathelin 结构域的高度保守的 N-末端前肽，分子质量通常在 12 ku 且含有单个二硫键[31]。大多数 cathelicidins 家族生物活性肽都含有四个半胱氨酸（C）残基，主要在中性粒细胞和黏膜等部位表达。成鱼体内 cathelicidins 的表达存在高度变异，并随 *cathelicidin* 基因和鱼体组织的变化而变化[32]。盲鳗科 cathelicidins 对革兰氏阴性菌和革兰氏阳性菌具有很强的抑菌活性，但对白色念珠菌（*Moniliaalbican*）无活性[33]。鳕科 cathelicidins 对革兰氏阴性菌及白色念珠菌具有一定的活性，但对革兰氏阳性菌无活性。另外，鲑科 cathelicidins 普遍存在一个包含六个氨基酸（RPGGGS）的重复基元，在遗传上具有不稳定性，呈现出与哺乳动物 cathelicidins 一定的相似性[34]。

2. 防御素　防御素（defensins）一般含有 6~8 个半胱氨酸，可以形成 β-折叠结构，分子质量为 3~6 ku[35]。防御素按照半胱氨酸的位置及配对情况划分，可以分为三个不同的构型：α-防御素、β-防御素和 θ-防御素。截至目前，仅在鱼类中发现一种构型的防御肽，即 β-防御素[36]。鱼 β-防御素是一类长度为 60~77 个氨基酸残基的防御肽[37]。除对细菌有抑菌活性外，鱼 β-防御素对鱼类特异性病毒，如病毒性出血性败血症病毒（VHSV）、病毒性神经坏死病毒（VNNV）和新加坡石斑鱼虹彩病毒（SGIV）等也具有抗病毒活性[38]。β-防御素的基因表达由 β-葡聚糖、脂多糖（LPS）和肽聚糖诱导，主要在皮肤中表达，也可在肾、鳃、脾等多种组织中高水平表达[39]。β-防御素在鱼类组织中的广泛分布证明了 β-防御素在先天免疫中的作用及其在对抗感染中的重要性[40]。目前，β-防御素除被用作抗菌剂外，也作为免疫应答因子，在细胞信号活动、未成熟树突状细胞激活以及免疫调节特性等方面发挥作用[41]。

3. Hepcidins　Hepcidins 是目前已知的含有最多半胱氨酸的阳离子生物活性肽[42]。Hepcidins 通常由 20~26 个氨基酸组成，结构特点为 4 个二硫键形成的稳定 β-折叠发夹结构[43]。Hepcidins 主要在肝脏中被分离，少部分在皮肤、鳃、肾脏、性腺、心脏和血液中表达[44]。目前，人们已经在鲑科、鲈科、鲽科及其他科中鉴定并分离出 30 多种 hepcidins。不同强度的病原体攻击、氧合水平和铁离子浓度是引起 hepcidins 表达变化的重要因素[45]。除了与运铁蛋白相互作用外，二硫键在维持 hepcidins 的稳定构象中起着至关重要的作用[46]。Hepcidins 具有很强的 DNA 结合能力，利用亮氨酸（L）在合成肽中取代半胱氨酸会导致 DNA 结合能力和抗菌活性的完全丧失[47]。Hepcidins 除了对原生动物寄生虫具有很强的活性外，还对革兰氏阴性菌和革兰氏阳性菌具有抗菌活性[48]。此外，hepcidins 还能够促进肠内铁吸收，促进巨噬细胞铁再循环。因此，hepcidins 也被称为铁调素[49,50]。

三、组蛋白衍生肽

组蛋白是核小体结构的主要组成部分，它通过形成组蛋白衍生肽在生物体的防御系统中起着至关重要的作用[51]。该肽已经在多种鱼类中被鉴定，并证明具有广谱抗微生物活性。除少部分组蛋白衍生肽在鳃、肠和脾脏等组织中发现外，其余均在皮肤中表达。最近，一种来自沃尔窄尾魟的组蛋白 H2A 衍生肽——hipposin 被鉴定并分离出来。同源性分析表明，hipposin 属于组蛋白 H2A 超家族，与鱼类中其他组蛋白来源的生物活性肽具有序列同源性。作为一种DNA结合蛋白，组蛋白是染色质的重要蛋白质组成部分，也是细菌和病毒蛋白的受体，主要与DNA稳定和基因表达调控有关[52]。因此，组蛋白衍生肽除具有抗菌、抗癌和抗生物膜等生物学活性外，还参与细胞的转录调控和凋亡，在机体先天免疫中起着关键作用[53]。

第二节 鱼源生物活性肽的生物学活性与作用机理

一、抗细菌

生物活性肽的抗细菌功能研究最为深入，几乎所有的鱼源生物活性肽对革兰氏阴性菌和革兰氏阳性菌都具有抗菌活性。对于革兰氏阴性菌，鲇皮肤黏液提取出的CF-14在最小抑菌浓度（MIC）为31.3 μg/mL时，表现出对大肠杆菌的抑菌活性[65]。伯氏肩孔南极鱼转录组获得的trematocine除对大肠杆菌具有抑菌活性外，还对短小芽孢杆菌（*Bacillus pumilus*）和枯草芽孢杆菌（*Psychrobacter sp.* TAD1）具有极强的抑菌作用[66]。翘嘴鲌肝脏中分离得到的leap-2在浓度为5×MIC条件时，1 h内即可杀死嗜水气单胞菌（*Aeromonas hydrophila*）[67]。NK-lysin截短得到的NKL-24通过膜活性细胞杀伤机制对副溶血弧菌（*Vibrio Parahaemolyticus*）具有良好的抗菌作用[68]。同样的结果在大黄鱼血红蛋白LCH4中被证实，在25 μg/mL的MIC浓度条件下LCH4即可抑制副溶血性弧菌的生长[69]。对于革兰氏阳性菌，最新从亚洲鲈中分离得到的moronecidins表现出良好的抑菌活性，在1×MIC浓度下作用4~6 h后表现出对鲍曼不动杆菌（*Acinetobacter baumannii*）的杀菌作用[70]。盲鳗肠道中提取的HF-18体外试验表明，在MIC浓度为4 μg/mL时即可产生对耐药金黄色葡萄球菌（*Staphylococcus aureus*）的抑菌作用，且在作用1 h内即可杀死细菌[71]。

通常情况下，鱼源生物活性肽对细菌的抗菌能力主要归功于其对细菌膜的破坏作用及形成孔洞的能力。带正电荷的生物活性肽与带有负电荷的细菌膜，通过调节线粒体的柔韧性，导致孔隙形成和线粒体解聚，最终导致细菌的死亡[72]。部分生物活性肽通过与DNA结合和竞争性抑制，穿透细菌膜并引起膜的扰动[65]。此外，组蛋白可以通过细胞膜穿透到细菌内，然后与其遗传物质结合或包裹病原体，形成中性粒细胞外网状陷阱（NETs），从而展现出广泛的抗菌功能[73]。然而，尽管组蛋白干预NETs的机制尚不清楚，但当组蛋白被特异性抗体阻断时，其抗菌功能也受到了抑制，这也证明了组蛋白在中性粒细胞外网状陷阱的形成中发挥了重要作用[74]。

二、抗病毒

与抗菌活性相比，鱼源生物活性肽的抗病毒活性的研究似乎还不够深入，但目前已经证明了鱼生物活性肽的抗病毒功能。最近发现，从美国红鱼TFPI-1肽C-末端截取的TO17具有抗病毒活性，将TO17与传染性脾肾坏死病毒（ISKNV）共孵育后注射到美国红鱼体内分别感染3、5和7 d后，美国红鱼脾脏内的病毒数量均有明显下降[75]。Guo等人[60]从斜带石斑鱼的肝脏中分离出β-防御素，能够减少新加坡石斑鱼虹彩病毒（SGIV）和神经坏死病毒（NNV）的感染和复制，就像EC-hepcidin-1和EC-hepcidin-2能够减少 SGIV 的发生一样[76]。此外，就 hepcidins 而言，在利用抗沙门氏菌弹状病毒（SCRV）和大嘴鲈沙门氏菌呼肠孤病毒（MsReV）对鲤上皮瘤细胞（EPC）和草鱼鳍细胞（GCF）进行感染时发现，金鼓鱼合成肽 SA-hepcidin-2 具有抗病毒活性，而 SA-hepcidin-1 则不具有抗病毒活性。

尽管已经证明了鱼源生物活性肽的抗病毒功能，但其机制仍不清楚。但最可能的原因是通过与病毒粒子直接结合、促使病毒颗粒凝集成团、抑制病毒增殖、利用病毒入侵并模仿病毒的侵染过程等实现其抗病毒作用[77]。

三、抗肿瘤

近年来，人们发现某些来源于鱼类的生物活性肽能够诱导肿瘤细胞的凋亡。Piscidins 对肿瘤细胞具有良好的抑制作用。Piscidin-4 可以通过激活外源性 Fas/FasL 受体途径介导 MG-63 细胞凋亡[78]。作为 α-微管蛋白的主要相互作用伴侣，生物活性肽还可以通过影响微管蛋白-微管平衡达到杀死肿瘤细胞的目的。例如，piscidin-4 可以与微管蛋白结合，破坏 A549 细胞中的微管网络，引起癌细胞中微管细胞骨架的缺陷[79]。最近，人们还发现斜带石斑鱼中鉴定出的 epinecidin-1 能够诱导线粒体超极化并产生 ROS，触发 caspase 依赖途径，造成恶性胶质母细胞瘤细胞 U87MG 的死亡和 DNA 损伤[16]。同样地，来源自杂交条纹鲈的 piscidin-1 在对正常细胞不产生副作用的同时，对骨肉瘤 OSA 细胞表现出更强的细胞毒性作用[80]。

四、其他活性

尽管研究较少，但鱼源生物活性肽对真菌和寄生虫也具有一定的抑制作用。Piscidin-2 对白色念珠菌（*Monilia albican*）和马拉色菌（*Pityrosporum*）具有有效的抗真菌作用，而来自斜带石斑鱼合成的 hepcidins 成熟肽 ecPi-2S、ecPi-3S 和 ecPi-4S 则具有较强的抗毕赤酵母（*Pichia pastoris*）活性[12]。关于抗寄生虫活性，Umasuthan 等人[81]证明了鲷分离的 Of-Pis1β 具有抗寄生虫活性。同样，piscidin-1 和 piscidin-2 及其衍生肽 piscidin-2β 在最低寄生浓度下显示出对梨状四膜虫（*T. pyriformis*）的抗寄生虫活性[82]。鱼源生物活性肽对寄生虫的抑制机理与其抑菌机理相似，主要是通过破坏寄生虫细胞膜，破坏膜电势平衡和干扰细胞正常代谢达到抗寄生虫的目的[83]。

第三节 鱼源生物活性肽的潜在功能及应用

目前，已有小部分新型生物活性肽被研发成功并应用于动物生产实践当中。将生物活性肽作为饲料添加剂使用，不仅可以提升动物的生长性能，还可以抑制动物肠道内病原菌的产生，调节动物机体微生态平衡。此外，充分了解并应用鱼源生物活性肽的其他潜在功能，将更好地促进我国水产及畜牧行业的健康发展。

一、免疫原性药物的开发

全世界对鱼卵和幼虫运输的需求日益增加，这不仅使得养殖业疾病的传播速度难以控制，也导致药物化合物的使用增加，从而导致细菌对抗生素等产生耐药性。目前，水产养殖业使用的疫苗并没有达到预期效果，这严重阻碍了渔业经济发展和动物福利。生物活性肽因其广泛的生物学活性和独特的作用模式，被认为是理想的水产养殖或临床开发抗病原微生物的工具。目前，已有相关研究关注鱼源生物活性肽作为免疫活性化合物的可能性。Pan等人[84]以杂交罗非鱼为研究对象，研究了TP3和TP4预处理、共处理和后处理杂交罗非鱼对创伤弧菌（*Vibrio vulnificus*）感染的保护作用。当病原菌与TP3和TP4共处理后，罗非鱼的存活率分别提高到95.3%和88.9%。此外，罗非鱼鱼腥蛋白样肽能够通过1型免疫应答（Th1）分子途径增强哺乳动物和鱼类的免疫应答[85]。

鱼源生物活性肽还可作为改进疫苗的潜在佐剂。目前，福尔马林灭活疫苗导致的严重过敏反应、皮疹、呼吸困难及肿胀等问题，亟须寻找降低疫苗灭活剂毒性的替代品。生物活性肽可以通过中和毒性因子或诱导细菌改变结构和形状，从而降低其毒性，这可能为灭活细菌病原体提供无害的手段。注射Epi-1灭活的日本脑炎病毒后，小鼠的存活率增加到100%。将pleurocidin（PLE）、哈维弧菌（*Vibrio harveyi*）抗原和重组甘油三磷酸脱氢酶（rGAPDH）蛋白包裹在聚丙交酯-乙交酯共聚物（PLG）中生产出的微粒（PLG-PLE/rGAPDH MP）能够稳定地释放PLE和rGAPDH，并且在两次疫苗接种后，对石斑鱼中的哈氏弧菌产生了长期的保护性免疫[86]。目前，基于生物活性肽的灭活疫苗在水产养殖中的应用报道还十分有限，但在未来，生物活性肽有望成为一种较好的福尔马林替代品。

二、抗氧化剂的开发

抗氧化化合物存在于每一种生物体内，可保护机体免受活性氧（ROS）对DNA、脂膜和蛋白质的损害。具有抗氧化性能的生物活性肽通常含有3~20个氨基酸残基，此类鱼源生物活性肽N-端区域通常含有疏水性氨基酸残基或酸性残基。通常条件下，生物活性肽不具有抗氧化特性，一旦其在胃肠道消化或发酵后，便释放出其抗氧化功能[12]。生物活性肽的功能往往取决于其结构、组成和具体序列。研究表明，富含半胱氨酸残基、硫化物键、疏水残基甚至组氨酸残基的阳离子短肽，往往被认为是具有超强抗氧化功能的候选物质。因此，某些来自鱼类的生物活性肽，可能比来自其他来源的肽具有更强的抗氧化效果。此外，尽管ROS造成的损害与许多影响公共健康的疾病有关，但适量的ROS能够影响DNA甚至脂质膜，增强生物活性肽的抗菌功能。虽然，鱼源生物活性肽具有成为抗氧化化合物的潜力，但生物活性肽作为抗氧化剂的潜在能力在鱼类中还未被深入探索。未来可在水产养殖、人类医学等领域将鱼源生物活性肽作为抗氧化化合物进行应用。

第四节 小结与展望

基于其自身广泛的生物学活性，鱼源生物活性肽在畜牧、食品以及医药领域均展现出巨大的潜在用途，但生物活性肽的相关研究仍存在一些问题，这些问题限制了它们在生产应用中的进一步发展。生物活性肽的局限性包含以下几点：一是肽的生产成本高，目前，绝大部分鱼源生物活性肽的获取仍是通过分离提取，发酵工程技术的广泛应用或许是解决此问题的有效途径。二是大部分天然生物活性肽细胞选择性差，现阶段常用的生物活性肽分子改造手段主要有氨基酸残基替换、肽链截取及活性中心杂合与结构优化。此外，也可利用脂肪酸修饰合成脂肽、运用纳米技术自组装或亲水性聚合形成纳米自组装肽或水凝胶肽，提高生物活性肽的细胞选择性。三是生物活性肽在体内易失活且对蛋白酶消化极具敏感性，目前，许多学者也在开发新策略以延长生物活性肽的半衰期，包括引入非天然氨基酸、肽链环化和 N-末端酰胺化等，防止生物活性肽在体内酶解。随着研究的不断深入，新发现的天然生物活性肽及人工合成的生物活性肽种类会越来越多，经济、高效、广谱的生物活性肽无疑具有巨大的应用潜力和广阔的发展前景。

参考文献

[1] 齐志涛，徐杨，邹钧，等．水产动物抗菌肽研究进展［J］．水产学报，2022，44（9）：1572-1583．

[2] Raju S V, Sarkar P, Kumar P, et al. Piscidin, fish antimicrobial peptide: structure, classification, properties, mechanism, gene regulation and therapeutical importance [J]. International Journal of Peptide Research and Therapeutics, 2020, 10: 1007.

[3] Zhuang Z R, Yang X D, Huang X Z, et al. Three new piscidins from orange-spotted grouper (*Epinephelus coioides*): phylogeny, expression and functional characterization [J]. Fish & Shellfish Immunology, 2017, 66: 240-253.

[4] Cetuk H, Maramba J, Britt M, et al. Differential interactions of piscidins with phospholipids and lipopolysaccharides at membrane interfaces [J]. Langmuir, 2020, 36 (18): 5065-5077.

[5] Mihailescu M, Sorci M, Seckute J, et al. Structure and function in antimicrobial piscidins: histidine position, directionality of membrane insertion, and pH-dependent permeabilization [J]. Journal of the American Chemical Society, 2019, 141 (25): 9837-9853.

[6] Cipolari O C, de Oliveira Neto X A, Conceicao K. Fish bioactive peptides: a systematic review focused on sting and skin [J]. Aquaculture, 2020, 515: 734598.

[7] Talandashti R, Mahdiuni H, Jafari M, et al. Molecular basis for membrane selectivity of antimicrobial peptide pleurocidin in the presence of different eukaryotic and prokaryotic model membranes [J]. Journal of Chemical Information and Modeling, 2019, 59 (7): 3262-3276.

[8] Hsu W H, Lai Y J, Wu S C. Effects of the anti-microbial peptide pardaxin plus sodium erythorbate dissolved in different gels on the quality of Pacific white shrimp under refrigerated storage [J]. Food Control, 2017, 73 (B): 712-719.

[9] Pan C Y, Lin C N, Chiou M T, et al. The antimicrobial peptide pardaxin exerts potent anti-tumor activity against canine perianal gland adenoma [J]. Oncotagret, 2015, 6 (4): 2290-2301.

[10] 尹航．非溶酶体途径脂质体用于干细胞转染和肿瘤细胞基因编辑的研究［D］．苏州：苏州大学，2019．

[11] 马萍．鲈鱼、石斑鱼、豹鳎鱼、泥鳅和臭蛙抗菌肽在毕赤酵母 SMD1168 中的高效表达［D］．兰州：兰州交通大学，2018．

[12] Valero Y, Saraiva-Fraga M, Costas B, et al. Antimicrobial peptides from fish: beyond the fight against pathogens [J]. Reviews in Aquaculture, 2020, 12 (1): 224-253.

[13] Browne M J, Feng C Y, Booth V, et al. Characterization and expression studies of Gaduscidin-1 and Gaduscidin-2; paralogous antimicrobial peptide-like transcripts from Atlantic cod (*Gadus*

morhua) [J]. Developmental & Comparative Immunology, 2011, 35 (3): 399-408.

[14] Sandhu G, Booth V, Morrow M R. Role of charge in lipid vesicle binding and vesicle surface saturation by gaduscidin-1 and gaduscidin-2 [J]. Langmuir, 2020, 36 (33): 9867-9877.

[15] Pan C Y, Chen J Y, Cheng Y S E, et al. Gene expression and localization of the epinecidin-1 antimicrobial peptide in the grouper (*Epinephelus coioides*), and its role in protecting fish against pathogenic infection [J]. DNA and Cell Biology, 2007, 26 (6): 403-413.

[16] Su B C, Wu T H, Hsu C H, et al. Distribution of positively charged amino acid residues in antimicrobial peptide epinecidin-1 is crucial for in vitro glioblastoma cytotoxicity and its underlying mechanisms [J]. Chemico-Biological Interactions, 2020, 315: 108904.

[17] Mason A J, Bertani P, Moulay G, et al. Membrane interaction of chrysophsin-1, a histidine-rich antimicrobial peptide from red sea bream [J]. Biochemistry, 2007, 46 (51): 15175-15187.

[18] Alexander T E, Smith I M, Lipsky Z W, et al. Role of lipopolysaccharides and lipoteichoic acids on C-Chrysophsin-1 interactions with model Gram-positive and Gram-negative bacterial membranes [J]. Biointerphases, 2020, 15 (0310073).

[19] Lauth X, Shike H, Burns J C, et al. Discovery and characterization of two isoforms of moronecidin, a novel antimicrobial peptide from hybrid striped bass [J]. The Journal of Biological Chemistry, 2002, 277 (7): 5030-5039.

[20] Campagna S, Saint N, Molle G, et al. Structure and mechanism of action of the antimicrobial peptide piscidin [J]. Biochemistry, 2007, 46 (7): 1771-1778.

[21] Silphaduang U, Noga E J. Peptide antibiotics in mast cells of fish [J]. Nature, 2001, 414 (6861): 268-269.

[22] Noga E J, Silphaduang U, Park N G, et al. Piscidin 4, a novel member of the piscidin family of antimicrobial peptides [J]. Comparative Biochemistry and Physiology B-Biochemistry & Molecular Biology, 2009, 152 (4): 299-305.

[23] Salger S A, Cassady K R, Reading B J, et al. A diverse family of host-defense peptides (piscidins) exhibit specialized anti-bacterial and anti-protozoal activities in fishes [J]. PloS One, 2016, 11 (e01594238).

[24] Salerno G, Parrinello N, Roch P, et al. cDNA sequence and tissue expression of an antimicrobial peptide, dicentracin; a new component of the moronecidin family isolated from head kidney leukocytes of sea bass, *Dicentrarchus labrax* [J]. Comparative Biochemistry and Physiology. Part B, Biochemistry & Molecular Biology, 2007, 146 (4): 521-529.

[25] Cole A M, Weis P, Diamond G. Isolation and characterization of pleurocidin, an antimicrobial peptide in the skin secretions of winter flounder [J]. The Journal of Biological Chemistry, 1997, 272 (18): 12008-12013.

[26] Douglas S E, Patrzykat A, Pytyck J, et al. Identification, structure and differential expression of novel pleurocidins clustered on the genome of the winter flounder, *Pseudopleuronectes americanus* (Walbaum) [J]. European Journal of Biochemistry, 2003, 270 (18): 3720-3730.

[27] Douglas S E, Gallant J W, Gong Z, et al. Cloning and developmental expression of a family of pleurocidin-like antimicrobial peptides from winter flounder, *Pleuronectes americanus* (Walbaum) [J]. Developmental & Comparative Immunology, 2001, 25 (2): 137-147.

[28] Sugiyama N, Araki M, Ishida M, et al. Further isolation and characterization of grammistins from the skin secretion of the soapfish *Grammistes sexlineatus* [J]. Toxicon, 2005, 45 (5): 595-601.

[29] Kaji T, Sugiyama N, Ishizaki S, et al. Molecular cloning of grammistins, peptide toxins from the soapfish *Pogonoperca punctata*, by hemolytic screening of a cDNA library [J]. Peptides, 2006, 27 (12): 3069-3076.

[30] Saitoh T, Seto Y, Fujikawa Y, et al. Distribution of three isoforms of antimicrobial peptide, chrysophsin-1, -2 and -3, in the red sea bream, *Pagrus major* (Chrysophrys) [J]. Analytical Biochemistry, 2019, 566: 13-15.

[31] Scheenstra M R, van Harten R M, Veldhuizen E J A, et al. Cathelicidins modulate TLR-activation and inflammation [J]. Frontiers in Immunology, 2020, 11: 1137.

[32] Bridle A, Nosworthy E, Polinski M, et al. Evidence of an antimicrobial-immunomodulatory role of Atlantic salmon cathelicidins during infection with *Yersinia ruckeri* [J]. PLoS One, 2011, 6 (8): e23417.

[33] Tomasinsig L, Zanetti M. The cathelicidins - structure, function and evolution [J]. Current Protein & Peptide Science, 2005, 6 (1): 23-34.

[34] 厉政. 细鳞鱼 Cathelicidins 的鉴定及宿主防御肽 CATH_BRALE 的结构与功能分析 [D]. 大连: 大连理工大学, 2013.

[35] Contreras G, Shirdel I, Braun M S, et al. Defensins: transcriptional regulation and function beyond antimicrobial activity [J]. Developmental & Comparative Immunology, 2020, 104: 103556.

[36] Ma Y Q, Kim S S, Maeng C H, et al. Key role of disulfide bridges in the antimicrobial activity of β-defensin from Olive Flounder [J]. International Journal of Peptide Research and Therapeutics, 2020, 26 (1): 291-299.

[37] Chaturvedi P, Dhanik M, Pande A. Molecular characterization and *in silico* analysis of defensin from *Tor putitora* (Hamilton) [J]. Probiotics and Antimicrobial Proteins, 2015, 7 (3): 207-215.

[38] Zhou Y C, Lei Y, Cao Z J, et al. A β-defensin gene of Trachinotus ovatus might be involved in the antimicrobial and antiviral immune response [J]. Developmental & Comparative Immunology, 2019, 92: 105-115.

[39] Harte A, Tian G, Xu Q, et al. Five subfamilies of β-defensin genes are present in salmonids: Evolutionary insights and expression analysis in Atlantic salmon *Salmo salar* [J]. Developmental & Comparative Immunology, 2020, 104: 103560.

[40] Yang K, Hou B R, Ren F F, et al. Characterization of grass carp (*Ctenopharyngodon idella*) beta-defensin 1: implications for its role in inflammation control [J]. Bioscience, biotechnology, and biochemistry, 2019, 83 (1): 87-94.

[41] Zhou Y, Zhou Q J, Qiao Y, et al. The host defense peptide β-defensin confers protection against *Vibrio anguillarum* in ayu, *Plecoglossus altivelis* [J]. Developmental & Comparative Immunology, 2020, 103: 103511.

[42] Lombardi L, Maisetta G, Batoni G, et al. Insights into the antimicrobial properties of hepcidins: advantages and drawbacks as potential therapeutic agents [J]. Molecules, 2015, 20 (4): 6319-6341.

[43] Shirdel I, Kalbassi M R, Hosseinkhani S, et al. Cloning, characterization and tissue-specific expression of the antimicrobial peptide hepcidin from caspian trout (*Salmo caspius*) and the antibacterial activity of the synthetic peptide [J]. Fish & Shellfish Immunology, 2019, 90: 288-296.

[44] Liu Y Y, Han X D, Chen X Z, et al. Molecular characterization and functional analysis of the hepcidin gene from rough skin sculpin (*Trachidermus fasciatus*) [J]. Fish & Shellfish Immunology,

2017, 68: 349-358.

[45] Xu T J, Sun Y N, Shi G, et al. Miiuy croaker hepcidin gene and comparative analyses reveal evidence for positive selection [J]. PLoS One, 2012, 7 (4): e35449.

[46] Chaturvedi P, Dhanik M, Pande A. Molecular characterization and in silico analysis of defensin from *Tor putitora* (Hamilton) [J]. Probiotics and Antimicrobial Proteins, 2015, 7 (3): 207-215.

[47] Hocquellet A, Le Senechal C, Garbay B. Importance of the disulfide bridges in the antibacterial activity of human hepcidin [J]. Peptides, 2012, 36 (2): 303-307.

[48] Xie J S, Obiefuna V, Hodgkinson J W, et al. Teleost antimicrobial peptide hepcidin contributes to host defense of goldfish (*Carassius auratus* L.) against *Trypanosoma carassii* [J]. Developmental & Comparative Immunology, 2019, 94: 11-15.

[49] Ma Y Q, Lee C J, Kim S S, et al. Role of hepcidins from black rockfish (*Sebastes schlegelii*) in iron-metabolic function and bacterial defense [J]. Journal of Marine Science and Engineering, 2020, 8 (7): 493.

[50] Jiang Y, Chen B, Yan Y L, et al. Hepcidin protects against iron overload-induced inhibition of bone formation in zebrafish [J]. Fish Physiology and Biochemistry, 2019, 45 (1): 365-374.

[51] 费威, 李慧, 马龙洋, 等. 水生动物组蛋白衍生抗菌肽的结构特征和抗菌活性研究进展 [J]. 河南农业科学, 2015, 44 (4): 9-13.

[52] Parseghian M H, Luhrs K A. Beyond the walls of the nucleus: the role of histones in cellular signaling and innate immunity [J]. Biochemistry and Cell Biology, 2006, 84 (4): 589-604.

[53] Athira P P, Anju M V, Anooja V V, et al. A histone H2A-derived antimicrobial peptide, hipposin from mangrove whip ray, *Himantura walga*: molecular and functional characterisation [J]. 3 Biotech, 2020, 10 (11): 467.

[54] Uzzell T, Stolzenberg E D, Shinnar A E, et al. Hagfish intestinal antimicrobial peptides are ancient cathelicidins [J]. Peptides, 2003, 24 (11): 1655-1667.

[55] Broekman D C, Frei D M, Gylfason G A, et al. Cod cathelicidin: Isolation of the mature peptide, cleavage site characterisation and developmental expression [J]. Developmental & Comparative Immunology, 2011, 35 (3): 296-303.

[56] Lu X J, Chen J, Huang Z A, et al. Identification and characterization of a novel cathelicidin from ayu, *Plecoglossus altivelis* [J]. Fish & Shellfish Immunology, 2011, 31 (1): 52-57.

[57] Dong J J, Wu F, Ye X, et al. Beta-defensin in Nile tilapia (*Oreochromis niloticus*): Sequence, tissue expression, and anti-bacterial activity of synthetic peptides [J]. Gene, 2015, 566 (1): 23-31.

[58] Wang G L, Li J H, Zou P F, et al. Expression pattern, promoter activity and bactericidal property of beta-defensin from the mandarin fish *Siniperca chuatsi* [J]. Fish & Shellfish Immunology, 2012, 33 (3): 522-531.

[59] Zou J, Mercier C, Koussounadis A, et al. Discovery of multiple beta-defensin like homologues in teleost fish [J]. Molecular Immunology, 2007, 44 (4): 638-647.

[60] Guo M L, Wei J G, Huang X H, et al. Antiviral effects of β-defensin derived from orange-spotted grouper (*Epinephelus coioides*) [J]. Fish & Shellfish Immunology, 2012, 32 (5): 828-838.

[61] Cuesta A, Meseguer J, Esteban M A. The antimicrobial peptide hepcidin exerts an important role in the innate immunity against bacteria in the bony fish gilthead seabream [J]. Molecular Immunology, 2008, 45 (8): 2333-2342.

[62] Pereiro P, Figueras A, Novoa B. A novel hepcidin-like in turbot (*Scophthalmus maximus* L.)

highly expressed after pathogen challenge but not after iron overload [J]. Fish & Shellfish Immunology, 2012, 32 (5): 879-889.

[63] Shike H, Shimizu C, Lauth X, et al. Organization and expression analysis of the zebrafish hepcidin gene, an antimicrobial peptide gene conserved among vertebrates [J]. Developmental & Comparative Immunology, 2004, 28 (7-8): 747-754.

[64] Lauth X, Babon J J, Stannard J A, et al. Bass hepcidin synthesis, solution structure, antimicrobial activities and synergism, and *in vivo* hepatic response to bacterial infections [J]. The Journal of Biological Chemistry, 2005, 280 (10): 9272-9282.

[65] Li T T, Liu Q W, Chen H T, et al. Antibacterial activity and mechanism of the cell-penetrating peptide CF-14 on the gram-negative bacteria, *Escherichia coli* [J]. Fish & Shellfish Immunology, 2020, 100: 489-495.

[66] Della Pelle G, Pera G, Belardinelli M C, et al. Trematocine, a novel antimicrobial peptide from the antarctic fish *Trematomus bernacchii*: identification and biological activity [J]. Antibiotics-Basel, 2020, 9 (2).

[67] Chen Y, Wu J, Cheng H L, et al. Anti-infective effects of a fish-derived antimicrobial peptide against drug-resistant bacteria and its synergistic effects with antibiotic [J]. Frontiers in Microbiology, 2020, 11: 602412.

[68] Shan Z G, Yang Y P, Guan N, et al. NKL-24: a novel antimicrobial peptide derived from zebrafish NK-lysin that inhibits bacterial growth and enhances resistance against *Vibrio parahaemolyticus* infection in *Yesso scallop*, *Patinopecten yessoensis* [J]. Fish & Shellfish Immunology, 2020, 106: 431-440.

[69] Yang S, Dong Y, Aweya J J, et al. A hemoglobin-derived antimicrobial peptide, LCH4, from the large yellow croaker (*Larimichthys crocea*) with potential use as a food preservative [J]. LWT-Food Science and Technology, 2020, 131: 109656.

[70] Taheri B, Mohammadi M, Nabipour I, et al. Identification of novel antimicrobial peptide from Asian sea bass (*Lates calcarifer*) by *in silico* and activity characterization [J]. PLoS One, 2018, 13 (10): e206578.

[71] Jiang M L, Yang X Q, Wu H M, et al. An active domain HF-18 derived from hagfish intestinal peptide effectively inhibited drug-resistant bacteria *in vitro/vivo* [J]. Biochemical Pharmacology, 2020, 172: 113746.

[72] Patel S, Akhtar N. Antimicrobial peptides (AMPs): the quintessential 'offense and defense' molecules are more than antimicrobials [J]. Biomedicine & Pharmacotherapy, 2017, 95: 1276-1283.

[73] Kawasaki H, Iwamuro S. Potential roles of histones in host defense as antimicrobial agents [J]. Infectious Disorders Drug Targets, 2008, 8 (3): 195-205.

[74] Wen L L, Zhao M L, Chi H, et al. Histones and chymotrypsin-like elastases play significant roles in the antimicrobial activity of tongue sole neutrophil extracellular traps [J]. Fish & Shellfish Immunology, 2018, 72: 470-476.

[75] He S W, Wang G H, Yue B, et al. TO17: A teleost antimicrobial peptide that induces degradation of bacterial nucleic acids and inhibits bacterial infection in red drum, *Sciaenops ocellatus* [J]. Fish & Shellfish Immunology, 2018, 72: 639-645.

[76] Zhou J G, Wei J G, Xu D, et al. Molecular cloning and characterization of two novel hepcidins from

orange‑spotted grouper, *Epinephelus coioides* [J]. Fish & Shellfish Immunology, 2011, 30 (2): 559‑568.

[77] 黄小建,李跃龙,戚南山,等.抗菌肽的生物学活性及其在畜禽生产中的作用[J].动物医学进展,2020,41 (12):115‑119.

[78] Kuo H M, Tseng C C, Chen N F, et al. MSP‑4, an antimicrobial peptide, induces apoptosis via activation of extrinsic Fas/FasL‑ and intrinsic mitochondria‑mediated pathways in one osteosarcoma cell line [J]. Marine Drugs, 2018, 16 (1).

[79] Ting C H, Liu Y C, Lyu P C, et al. Nile Tilapia derived antimicrobial peptide TP4 exerts antineoplastic activity through microtubule disruption [J]. Marine Drugs, 2018, 16 (12).

[80] Cheng M H, Pan C Y, Chen N F, et al. Piscidin‑1 induces apoptosis via mitochondrial reactive oxygen species‑regulated mitochondrial dysfunction in human osteosarcoma cells [J]. Scientific Reports, 2020, 10 (1): 5045.

[81] Umasuthan N, Mothishri M S, Thulasitha W S, et al. Molecular, genomic, and expressional delineation of a piscidin from rock bream (*Oplegnathus fasciatus*) with evidence for the potent antimicrobial activities of Of‑Pis1 peptide [J]. Fish & Shellfish Immunology, 2016, 48: 154‑168.

[82] Ruangsri J, Salger S A, Caipang C M A, et al. Differential expression and biological activity of two piscidin paralogues and a novel splice variant in Atlantic cod (*Gadus morhua* L.) [J]. Fish & Shellfish Immunology, 2012, 32 (3): 396‑406.

[83] 王雪洋,韩淑敏,李金库,等.抗菌肽在畜禽生产中的应用进展[J].畜牧与饲料科学,2019,40 (8):35‑37.

[84] Pan C Y, Tsai T Y, Su B C, et al. Study of the antimicrobial activity of Tilapia piscidin 3 (TP3) and TP4 and their effects on immune functions in hybrid Tilapia (*Oreochromis* spp.) [J]. PLoS One, 2017, 12 (1): e169678.

[85] Acosta J, Carpio Y, Valdés I, et al. Co‑administration of tilapia alpha‑helical antimicrobial peptides with subunit antigens boost immunogenicity in mice and tilapia (*Oreochromis niloticus*) [J]. Vaccine, 2014, 32 (2): 223‑229.

[86] Liu S P, Chuang S C, Yang C D. Protective immunity against *Vibrio harveyi* in grouper induced by single vaccination with poly (Lactide‑co‑glycolide) microparticles releasing pleurocidin peptide and recombinant glyceraldehyde‑3‑phosphate dehydrogenase [J]. Vaccines, 2020, 8 (1): 33.

第四篇
氨基酸的应用与检测

第十六章 水产动物氨基酸的应用与检测

第一节 氨基酸的平衡和互补作用

一、氨基酸的平衡

所谓氨基酸的平衡是指饲料可利用的各种必需氨基酸的组成和比例与动物对必需氨基酸的需要相同或非常相近。当饲料中所含有的可利用的必需氨基酸处于平衡状态时，才能获得理想的蛋白质效率。如果氨基酸不平衡，即使蛋白质含量很高，也不能获得高的蛋白质效率。氨基酸平衡犹如木桶盛水的道理，当氨基酸不平衡时，好比一个木桶的桶板长短不一，盛水容积小。饲料氨基酸的平衡是衡量饲料蛋白质质量的最重要指标。

蛋白质的互补作用（又称氨基酸的互补作用）是指利用不同蛋白源的氨基酸组成特点，相互取长补短使饲料的氨基酸趋于平衡。在生产实践中，这是提高蛋白质利用率最为经济、有效的方法[1]。

二、氨基酸的不平衡

饲料氨基酸的比例与动物氨基酸需要之间的比例一致程度很差，氨基酸不平衡，饲料营养价值低。氨基酸不平衡主要是比例问题，而氨基酸缺乏主要是数量不足。实际生产中，饲料氨基酸不平衡一般同时存在氨基酸缺乏。一般不会出现饲料中所有氨基酸的比例都超过需要的情况，经常表现为个别氨基酸比例偏高，少数氨基酸比例偏低。通过饲料原料的搭配和补充合成氨基酸，可以改善饲料氨基酸平衡状况，提高动物生产性能。

三、氨基酸的互补

不同饲料原料的氨基酸含量和比例上存在很大差异，通过两种或多种饲料原料的搭配，可以取长补短，弥补个别原料氨基酸组成上的缺陷，改善饲料氨基酸平衡，提高总体蛋白质的营养价值[2]。

四、理想蛋白质

理想蛋白质是指氨基酸组成和比例与动物氨基酸需要完全一致的蛋白质。理想蛋白质不但必需氨基酸比例完全平衡，而且必需氨基酸和非必需氨基酸之间也完全平衡。动物对理想蛋白质的利用率为100%。理想蛋白质的本质是氨基酸间的最佳平衡模式，以这种模式组成的饲料蛋白质最符合动物的需要,因而能够最大限度地被利用。在理想蛋白质模式中所有氨基酸都被看作必需和同等重要的,增加或减少任何一种氨基酸都会破坏这种最佳平衡状态,降低氨基酸的利用效率[1]。

理想蛋白质中最重要的是必需氨基酸的比例,为了便于推广应用,通常把赖氨酸作为基准氨基酸,其相对需要量定位 100,其他必需氨基酸需要量表示为与赖氨酸需要量的百分比,称为必需氨基酸模式或理想蛋白模式[3]。由于生产方式、代谢途径的差异,不同种类和不同生产阶段的动物理想蛋白模式存在明显差异。应用理想蛋白质模式要考虑实际饲料氨基酸的消化率,以可消化氨基酸为基础的理想蛋白质模式可以使动物对氨基酸的需要达到最大的满足和最小的浪费。

第二节 水产动物氨基酸的有效性

一、晶体氨基酸的利用

水产饲料中添加晶体氨基酸的效果并不理想,并因种类的不同而不同,甚至存在一些相反的报道。比较一致的看法是,在鲑鳟饲料中添加晶体氨基酸较为有效,而在虾类及无胃的鲤科鱼类饲料中添加晶体氨基酸则无效或效果不明显。一些研究表明,海水鱼类可有效利用饲料中的晶体氨基酸,如大菱鲆幼鱼饲料中 19% 的完整蛋白可为晶体氨基酸代替而不会对鱼体生长产生不利影响[4]。水产饲料中添加晶体氨基酸的作用效果不佳的原因主要包括以下几个方面:

1. 氨基酸的吸收不同步 饲料中添加的晶体氨基酸在进入消化道后很快被鱼体吸收进入血液、组织液中,在血液或组织液中很快形成单个氨基酸的高峰值。晶体氨基酸吸收过快,不能和饲料中蛋白态的结合氨基酸同步吸收,并且从血液中移出的速度比结合氨基酸快,从而使得组织中氨基酸不平衡,使用来合成蛋白质的氨基酸减少。因此,相对饲喂蛋白态结合氨基酸而言,饲喂晶体氨基酸的鱼类其血浆游离氨基酸高峰值不仅时间提前,而且峰值也显著提高。由于氨基酸合成蛋白质必须按一定比例方可进行,且水生动物储存游离氨基酸的能力甚低,导致先吸收的晶体氨基酸不能用于合成蛋白质(从蛋白质来源的氨基酸尚未被吸收)而直接排泄或代谢。此外,添加的游离氨基酸还影响其他必需氨基酸吸收的同步化,使得氨基酸间得不到平衡互补,从而影响游离氨基酸乃至整个蛋白质的利用率[5]。

氨基酸被不同步吸收入体内后,对于不符合比例要求的氨基酸,其命运通常是通过鳃、肾排泄或被氧化分解。Murai 等[6]发现,鲤摄食以晶体氨基酸作为蛋白源的饲料后,24 h 内有 36% 的氨基酸通过鳃或肾排入养殖水体;摄食晶体氨基酸和酪蛋白混合物作为蛋白源的饲料,或以酪蛋白、明胶为蛋白源的饲料,其排入水体中的氨基酸分别为 12.8%、1%。由上可见,由于氨基酸吸收不同步造成的氨基酸浪费(未用于构建机体组织)是巨大的,这是水产饲料中添加晶体氨基酸效果不佳的重要原因。

2. 氨基酸的水中溶失 在饲料中经常添加的晶体氨基酸有 L-赖氨酸盐酸盐(硫酸盐)、DL-蛋氨酸、L-苏氨酸和 L-色氨酸等。这些氨基酸都具有水溶性,在饲料中直接添加则易溶失于水中,而导致利用率降低。刘永坚等[7]在实用基础饲料中分别添加 0.4%

晶体赖氨酸或包膜赖氨酸，取 10 g 饲料置于 100 mL 水中 10 min，晶体赖氨酸的溶失率为 $(13.22\pm3.5)\%$，包膜赖氨酸溶失率为 $(4.81\pm0.8)\%$。由此可见，晶体氨基酸的水中溶失是客观存在的。对于抱食咀嚼的虾蟹类、摄食缓慢的鱼类，饲料中添加的晶体氨基酸在水中的溶失更为显著，再考虑到生产中养殖水体的水流、搅动等因素，更加剧了晶体氨基酸的溶失。

3. 养殖品种 添加晶体氨基酸的作用效果还与养殖品种及其消化道结构特点有一定关系。一般而言，冷水性鱼类（如虹鳟）能够有效利用晶体氨基酸，而温水性鱼类（如鲤）却不能。研究发现，冷水性鱼类对饲料氨基酸的吸收速度显著低于温水性鱼类。Walton 和 Wilson[8]报道，采食饲料 48 h 内虹鳟肝脏中游离氨基酸的浓度相对比较稳定，并不随饲料氨基酸浓度变化而变化；而鲤采食饲料后肝脏游离氨基酸浓度随饲料氨基酸含量呈正相关变化。

有胃鱼与无胃鱼对晶体氨基酸的利用不同。无胃鱼肠道相对细短，新陈代谢速度快，且其消化功能同水温关系极大，摄食季节性很强，因此无胃鱼类能否有效利用晶体氨基酸一直存在争论。普遍接受的观点是在饲料中添加晶体氨基酸不能为无胃鱼类同步吸收利用，从而达不到理想的促生长效果。陈丙爱[9]比较了鲤、罗非鱼对饲料中添加晶体氨基酸的利用效果，发现添加晶体氨基酸对鲤的生长性能基本无改善，但显著促进了罗非鱼的生长；二者摄食后的血清游离氨基酸浓度变化也表现出不同的特点。出现这种差异的一个重要事实是鲤属无胃鱼类，而罗非鱼则具有一个较为发达的胃。胃的存在，对食物具有储存和向肠道缓慢释放的作用，因而在一定程度上缓解了晶体氨基酸与蛋白态氨基酸吸收不同步的矛盾。

二、提高晶体氨基酸的利用

可见，晶体氨基酸作用效果不佳的原因主要是由晶体氨基酸吸收速度过快导致的吸收不同步和水中溶失造成的。针对这两种原因，可以采取提高投饲频率和对氨基酸进行缓释处理等方法予以解决。

1. 提高投饲频率 投饲频率是影响晶体氨基酸作用效果的重要因素。提高投饲频率，缩短投饲间隔后，可以使前次投饲时所产生的血液氨基酸依然保持在一个较高的水平，后一次投饲产生的血液氨基酸峰值可与前次产生一定程度的叠加，使晶体氨基酸与其他来源于饲料蛋白态的氨基酸产生一定互补，从而改善晶体氨基酸的作用效果[5]。

2. 包膜氨基酸 为减少晶体氨基酸在水中的溶失，延缓晶体氨基酸在肠道中的吸收速度，可对晶体氨基酸进行缓释处理，采用成膜材料把固体或液体包覆形成微小粒子，使被包被物质缓慢释放。氨基酸经包膜后，在消化道中需先经过包膜材料的崩解才可被吸收，因而其吸收速度降低，可在一定程度上缓解游离氨基酸与蛋白态氨基酸吸收不同步的矛盾。刘永坚等[7]在实用饲料中添加 0.4%赖氨酸，包膜赖氨酸组草鱼的增重率从 192.3%提高到 222.0%，而添加 0.4%晶体赖氨酸则无改善。在日本对虾基础饲料（对照组，不添加晶体氨基酸）中补充晶体或包膜赖氨酸、蛋氨酸（1.45%＋1.21%），养殖 42 d 后对照组、补充晶体氨基酸和补充包膜氨基酸的增重率分别为 194%、204%和 271%，显示出添加包膜氨基酸对生长的显著改善效应[10]。

第三节 氨基酸之间的交互作用

水产动物饲料氨基酸之间的相互关系主要是指颉颃作用和协同作用。

一、协同作用

协同作用是指增加饲料中一种氨基酸的添加量而导致另外一种或几种氨基酸需要量下降的现象。

1. 含硫氨基酸 含硫氨基酸主要是指分子结构中含有硫的氨基酸，包括蛋氨酸、半胱氨酸以及牛磺酸等。由于其在代谢途径中的相关性以及存在蛋氨酸向其他含硫氨基酸转化的能力（图16-1）。所以人们通常将含硫氨基酸作为一个整体来研究其功能和需要量。在对美国红鱼饲料中添加不同比例胱氨酸和蛋氨酸的研究中发现，饲料中的胱氨酸对蛋氨酸具有协同作用，可以节约50%的蛋氨酸。即在足量的胱氨酸条件下，美国红鱼对蛋氨酸的需要量可以减少50%[11]。在对虹鳟饲料中牛磺酸和蛋氨酸交互作用的研究中，研究人员应用3个水平的牛磺酸和3个水平的蛋氨酸进行双因素试验，在以生长性能为评价指标时，二者交互作用不显著，但是在对血清蛋氨酸含量、血清胰岛素样生长因子IGF-I等指标进行评价时，饲料中牛磺酸和蛋氨酸的交互作用显著[12]。

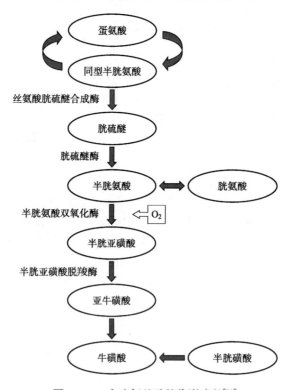

图16-1 含硫氨基酸的代谢途径[13]

2. 苯丙氨酸和酪氨酸 由于其结构的特点，它们之间存在着协同作用。苯丙氨酸可以转化为酪氨酸，而酪氨酸却不能转化为苯丙氨酸。如果饲料中酪氨酸足够，可以减少苯丙氨酸的用量，从而节约这一氨基酸。

二、颉颃作用

氨基酸的颉颃，指的是饲料中某一种或几种氨基酸的浓度过高情况下，影响其他氨基酸的吸收和利用，降低氨基酸的利用率。氨基酸颉颃的作用机理主要是肠道吸收过程中转运载体的竞争、肾小管重吸收过程中转运载体的竞争、代谢过程中相关酶活性的变化等。

1. 赖氨酸与精氨酸，组氨酸与精氨酸 这些氨基酸均属于碱性氨基酸，一方面两者具有相同的肠道吸收途径和肾小管重吸收途径，饲料赖氨酸过高则妨碍精氨酸在肠道的吸收和肾小管的重吸收，导致尿中排出的精氨酸量增加；另一方面是赖氨酸能导致肾脏线粒体中精氨酸酶活性升高，加快精氨酸降解，造成精氨酸缺乏。此外，过高的精氨酸水平会导致赖氨酸-酮戊二酸还原酶活性增加，从而增加赖氨酸的需要量。一般认为在高等陆生动物中，赖氨酸对精氨酸有颉颃作用，过量的赖氨酸会影响精氨酸的吸收和重吸收。在水生动物中的情况有些复杂。只要在赖氨酸和精氨酸二者都足量的情况下，过量的赖氨酸或精氨酸不会抑制牙鲆幼鱼的生长，并且血浆中精氨酸、鸟氨酸和尿素浓度都没有显著差异，肝脏精氨酸酶活力也没有显著变化[14]。但饲料中赖氨酸含量升高，大西洋鲑血浆中精氨酸、鸟氨酸和尿素含量同时下降。大西洋鲑对赖氨酸和精氨酸吸收的离体实验研究表明，较低浓度（1 mmol/L）的赖氨酸并没有影响精氨酸的积累，但赖氨酸的浓度增大到 3.4 mmol/L 时，精氨酸的积累就减少了 60%；另一方面在培养液中添加 1 mmol/L 的精氨酸就可降低赖氨酸的积累[15]。可见精氨酸和赖氨酸可能有共同的载体，并且精氨酸与载体的亲和力比赖氨酸更强。

精氨酸与组氨酸同属于碱性氨基酸，二者在鱼体内的代谢途径具有相关性，且具有相同的代谢终产物。以不同比例的精氨酸和组氨酸饲喂牙鲆幼鱼 70 d 后，通过比较生长、血液指标、抗氧化能力及应激反应后发现，两种氨基酸存在显著的颉颃作用[16]。

2. 支链氨基酸之间 亮氨酸、异亮氨酸和缬氨酸这三种氨基酸都是支链氨基酸，结构非常类似，在肠道吸收、肾小管重吸收过程中存在竞争。实际生产中过量的亮氨酸严重抑制采食和生长，可通过额外添加异亮氨酸和缬氨酸得到缓解；反之，过量异亮氨酸和缬氨酸的生长抑制作用也可通过添加更多的亮氨酸缓解。以高添加量的亮氨酸与缬氨酸饲喂牙鲆发现，两种支链氨基酸表现出显著的颉颃作用。但是在其中一种支链氨基酸不足时，增加另一种支链氨基酸的添加量可以提高牙鲆幼鱼的生长性能[17]。但 Choo 等[18]对虹鳟的研究认为，过量的亮氨酸抑制生长是由于其过量产生的毒性而非亮氨酸对其他支链氨基酸的颉颃作用。出现这种不同可能是由于以上研究主要以添加晶体氨基酸的方式调节饲料中氨基酸的比例，养殖鱼类在吸收利用晶体氨基酸时会有与其他蛋白源吸收利用不同步的现象，而这种不同步性造成了各项研究结果的不一致。

第四节 水产动物氨基酸与营养性疾病

一、氨基酸的缺乏

一种或几种必需氨基酸含量不能满足动物需要的情况称为氨基酸缺乏。氨基酸缺乏时，动物只能以所缺乏的氨基酸满足蛋白质合成的程度来利用其他氨基酸，产生蛋白质缺乏症，降低动物生产性能，同时多余的其他氨基酸被氧化供能转化为其他物质，降低能量利用率，增加肝和肾的负担，增加热应激。氨基酸缺乏一般在蛋白质水平低的情况下容易出现，可通过补充相应氨基酸得以预防。

必需氨基酸的缺乏相当于蛋白质的缺乏，必需氨基酸缺乏与蛋白质缺乏表现症状相同。例如，幼龄动物食欲减退或废绝，成年动物采食量代偿性增加。体内酶和激素合成速度减慢，血清蛋白浓度降低；生长速度下降，饲料利用率变差，出现负氮平衡。但氨基酸的缺乏也不完全等同于蛋白质缺乏，不同必需氨基酸的缺乏症不尽相同，某些氨基酸缺乏还可能产生其特异性缺乏症。

二、氨基酸的中毒

氨基酸过量表现出毒性作用，不同种类氨基酸毒性差异很大。蛋氨酸的毒性远远大于其他氨基酸。氨基酸的毒性可能与该氨基酸与其他氨基酸之间的颉颃作用有关。当某一氨基酸过量时，可能导致其他某些氨基酸利用率下降，从而导致这些氨基酸的缺乏。自然条件下几乎不存在氨基酸中毒，常发生在大量使用合成氨基酸时。

第五节　氨基酸的检测

1. 甲醛滴定法　甲醛滴定法的原理是向中性或碱性氨基酸溶液中加入甲醛，氨基酸的氨基和甲醛可形成羟甲基衍生物，使用碱标准溶液滴定该衍生物，采用酚酞指示剂确定滴定终点。实验步骤通常是以甲酸溶解样品，加入冰乙酸作为滴定介质，以高氯酸标准溶液滴定并计算得到主成分氨基酸的含量。此法在氨基酸类饲料添加剂主成分测定中应用较多，标准《饲料添加剂 L-色氨酸》（GB/T 25735—2010）、《饲料级 L-赖氨酸盐酸盐》（NY 39—1987）均采用此法分别对产品中的主成分指标 L-苏氨酸、L-色氨酸以及 L-赖氨酸盐酸盐的含量进行定量分析[19]。

2. 碘量法　碘量法是以碘或碘化物作为氧化剂或还原剂进行滴定的方法，在实际实验中分为直接碘量法和间接碘量法。标准《饲料级 DL-蛋氨酸》（GB/T 17810—2009）采用间接碘量法原理测定饲料添加剂 DL-蛋氨酸，以碘与硫代硫酸钠的特征反应作为方法原理，以磷酸盐缓冲液作为溶剂调节控制反应溶液的 pH，淀粉溶液作为指示剂对滴定终点进行判断[19]。

3. 紫外-可见分光光度法　紫外-可见分光光度法是通过测定 190~800 nm 波长范围内待测物质的吸光值，对化合物进行定量分析的方法。部分氨基酸如酪氨酸、色氨酸、苯丙氨酸等因其结构中含有苯环共轭双键系统，因而在紫外光区存在明显特征吸收，对于在紫外光区没有特征吸收的氨基酸则需要进行衍生化反应，生成具有紫外吸收的化合物，之后采用紫外分光光度计可测得对应氨基酸组分的含量。《饲料中色氨酸的测定》（GB/T 15400—2018）中包含分光光度法测定饲料中色氨酸的含量。饲料样品中蛋白经碱水解后，在酸性介质、氧化剂存在的条件下，色氨酸吲哚环与对二甲氨基苯甲醛反应生成蓝色化合物，以紫外分光光度法在 590 nm 处测定其特征吸收，通过绘制标准曲线测定色氨酸含量[19]。

4. 离子交换色谱法 离子交换色谱法是根据离子交换原理结合液相色谱技术对溶液中的阳离子和阴离子进行分离分析的方法，广泛应用于无机阴、阳离子以及氨基酸、有机酸、核酸等有机物的分离分析中。此法利用氨基酸在酸性条件下形成阳离子而通过阳离子交换色谱柱进行分离，之后经过茚三酮柱后衍生、紫外-可见光检测器进行测定。

市售的自动氨基酸分析仪多采用高效阳离子交换色谱-柱后茚三酮衍生光度检测技术。这种方法准确可靠、重现性好，能测定大多数种类的氨基酸及其同系物。标准《饲料中氨基酸的测定》(GB/T 18246—2019)、《饲料中含硫氨基酸的测定 离子交换色谱法》(GB/T 15399—2018) 中对饲料中氨基酸的测定步骤进行了详细规范，根据待测目标化合物的不同，分别以盐酸水解法、过甲酸氧化水解法以及碱水解法对饲料样品进行前处理，采用氨基酸自动分析仪进行测定。

5. 高效液相色谱法 该方法包括柱前衍生和柱后衍生两大类。目前多用柱前衍生-反相高效液相色谱法，即首先将氨基酸转化为适于反相液相色谱分离并能被灵敏检测的衍生物，然后采用高效液相色谱法对上述衍生物进行分离和检测。该方法灵敏、快速、应用范围广、易于自动化[20]。

6. 气相色谱法 采用气相色谱法（GC）分离测定氨基酸多需要进行柱前衍生化反应，将氨基酸衍生为弱极性、挥发性和热稳定性好的物质后再进行仪器测定，衍生化反应包括酯化-酰化反应、烷基化反应、硅烷化反应等，常用的衍生剂包括氯甲酸甲酯（MCF）、甲基三甲基甲硅烷基三氟乙酰胺（MSTFA）、六甲基二硅氮烷（MSDS）等。此法的优点在于分离效率高、柱效高、适于氨基酸手性拆分以及易于与质谱联用，缺点是衍生产物多，容易干扰氨基酸组分的测定，专一性差[19]。

7. 近红外光谱法 近年来，近红外光谱技术因其快速分析、无损检测、无污染、重现性好等优点，在食品、农产品、饲料等多个检测领域的研究得到了迅速发展。近红外光谱法的原理是利用有机物质的含氢基团（C—H、N—H、O—H等）在波长780～2 526 nm的电磁波区域跃迁时产生的光谱变化，结合计算机与化学计量学进行样品中多种成分含量的测定。主要步骤包括样品光谱采集、光谱预处理、建立定标模型以及外部验证等。采用近红外光谱法可对各种饲料原料中氨基酸、蛋白质以及水分进行快速测定[19]。

参考文献

[1] 麦康森. 水产动物营养与饲料学 [M]. 2版. 北京：中国农业出版社，2011：10-28.

[2] 陈代文，余冰. 动物营养学 [M]. 北京：中国农业出版社，2020

[3] 刁其玉. 动物氨基酸营养与饲料 [M]. 北京：化学工业出版社，2007.

[4] Helena P, Aires O. The effect of dietary protein replacement by crystalline amino acid on growth and nitrogen utilization of turbot *Scophthalmus maximus* juveniles [J]. Aquaculture, 2005, 250：755-764.

[5] 冷向军，李小勤，陈丙爱，等. 鱼类对晶体氨基酸利用的研究进展 [J]. 水生生物学报，2009，33(1)：119-123.

[6] Murai T, Ogata H, Takeuchi T, et al. Composition of free amino acid in excretion of carp fed amino acid diets and casein-gelatin diets [J]. Bulletin of the Japanese Society of Scientific Fisheries, 1984,

50 (11)：1957.
- [7] 刘永坚，田丽霞，刘栋辉，等．实用饲料补充结晶或包膜氨基酸对草鱼生长、血清游离氨基酸肌肉蛋白质合成率的影响［J］．水产学报，2002，26（3）：252-258.
- [8] Walton M J, Wilson R P. Postprandial changes in plasma and liver free amino acids of rainbow trout fed complete diets containing casein［J］. Aquaculture, 1986，51：105-115.
- [9] 陈丙爱．鲤鱼、罗非鱼对晶体氨基酸利用的比较研究［D］．上海：上海水产大学，2007.
- [10] Alam M S, Teshima S, Koshio S, et al. Supplemental effects of coated methionine and/or lysine to soy protein isolate diet for juvenile kuruma shrimp, *Arsupenaeus japonicus*［J］. Aquaculture, 2005，248：13-19.
- [11] Goff J B, Gatlin D M. Evaluation of different sulfur amino acid compounds in the diet of red drum, *Sciaenops ocellatus*, and sparing value of cystine for methionine［J］. Aquaculture, 2004，241（1-4）：465-477.
- [12] Gaylord T G, Barrows F T, Teague A M, et al. Supplementation of taurine and methionine to all-plant protein diets for rainbow trout（*Oncorhynchus mykiss*）［J］. Aquaculture, 2007，269（1-4）：514-524.
- [13] 韩雨哲，高乔，崔培，等．水产动物饲料中氨基酸交互作用的研究［J］．天津农学院学报，2015，22（1）：45-49.
- [14] Shah A M, Shin-Ichi T, Manabu I, et al. Effects of dietary arginine and lysine levels on growth performance and biochemical parameters of juvenile Japanese flounder *Paralichthys olivaceus*［J］. Fisheries Science, 2002，68（3）：509-606.
- [15] Berge G, Bakke-McKellep A M, Lied E, . In vitro uptake and interaction between arginine and lysine in the intestine of Atlantic salmon（*Salmo salar*）［J］. Aquaculture, 1999，179：（1-4）：181-193.
- [16] Han Y, Koshio S, Ishikawa M, et al. Interactive effects of dietary arginine and histidine on the performances of Japanese flounder *Paralichthys olivaceus* juveniles［J］. Aquaculture, 2013，（414-415）：173-182.
- [17] Han Y, Han R, Koshio S, et al. Interactive effects of dietary valine and leucine on two sizes of Japanese flounder *Paralichthys olivaceus*［J］. Aquaculture, 2014，432：130-138.
- [18] Choo P S, Smith T K, Cho C Y, et al. Dietary excesses of leucine influence growth and body-composition of rainbow-trout［J］. The Journal of Nutrition, 1991，121（12）：1932-1939.
- [19] 贾铮，李兰，赵根龙，等．饲料中氨基酸分析技术研究进展［J］．农产品质量与安全，2017，6：3-60.
- [20] 江海风，马品一，金月，等．氨基酸分析方法的研究进展［J］．现代科学仪器，2013，4：55-61.

附 录

附表一 水产动物的氨基酸营养需要

淡水鱼类的氨基酸营养需要（以干物质计,%）

项目	大西洋鲑	鲤	南亚野鲮	罗非鱼	斑点叉尾鲴	杂交条纹鲈	虹鳟	太平洋鲑
可消化蛋白质	36	32	32	29	29	36	38	40
精氨酸	1.8	1.7	1.7	1.2	1.2	1.0	1.5	2.2
组氨酸	0.8	0.5	0.9	1.0	0.6	NT	0.8	0.7
异亮氨酸	1.1	1.0	1.0	1.0	0.8	NT	1.1	1.0
亮氨酸	1.5	1.4	1.5	1.9	1.3	NT	1.5	1.6
赖氨酸	2.4	2.2	2.3	1.6	1.6	1.6	2.4	2.2
蛋氨酸	0.7	0.7	0.7	0.7	0.6	0.7	0.7	0.7
蛋氨酸+半胱氨酸	1.1	1.0	1.0	1.0	0.9	1.1	1.1	1.1
苯丙氨酸	0.9	1.3	0.9	1.1	0.7	0.9	0.9	0.9
苯丙氨酸+酪氨酸	1.8	2.0	1.6	1.6	1.6	NT	1.8	1.8
苏氨酸	1.1	1.5	1.7	1.1	0.7	0.9	1.1	1.1
色氨酸	0.3	0.3	0.4	0.3	0.2	0.3	0.3	0.3
缬氨酸	1.2	1.4	1.5	1.5	0.8	NT	1.2	1.2

注：NT，未检测。

海水鱼类的氨基酸营养需要（以干物质计,%）

项目	尖吻鲈	军曹鱼	欧洲狼鲈	褐牙鲆	石斑鱼	美国红鱼	黄条鰤
可消化蛋白质	38	38	40	40	42	36	38
精氨酸	1.8	NT	1.8	2.0	NT	1.8	1.6
组氨酸	NT	NT	NT	NT	NT	NT	NT
异亮氨酸	NT	NT	NT	NT	NT	NT	NT
亮氨酸	NT	NT	NT	NT	NT	NT	NT
赖氨酸	2.1	2.3	2.2	2.6	2.8	1.7	1.9
蛋氨酸	0.8	0.8	NT	0.9	NT	0.8	0.8

(续)

项目	尖吻鲈	军曹鱼	欧洲狼鲈	褐牙鲆	石斑鱼	美国红鱼	黄条鰤
蛋氨酸＋半胱氨酸	1.2	1.1	1.1	NT	NT	1.2	1.2
苯丙氨酸	NT	NT	NT	NT	NT	NT	NT
苯丙氨酸＋酪氨酸	NT	NT	NT	NT	NT	NT	NT
苏氨酸	NT	NT	1.2	NT	NT	0.8	NT
色氨酸	NT	NT	0.3	NT	NT	NT	NT
缬氨酸	NT	NT	NT	NT	NT	NT	NT

注：NT，未检测。

海水虾类的氨基酸营养需要（以干物质计，%）

项目	日本囊对虾	凡纳滨对虾	斑节对虾
可消化蛋白质	38	30	34
精氨酸	1.6	NT	1.9
组氨酸	0.6	NT	0.8
异亮氨酸	1.3	NT	1.0
亮氨酸	1.9	NT	1.7
赖氨酸	1.9	1.6	2.1
蛋氨酸	0.7	NT	0.7
蛋氨酸＋半胱氨酸	1.0	NT	1.0
苯丙氨酸	1.5	NT	1.4
苯丙氨酸＋酪氨酸	NT	NT	NT
苏氨酸	1.3	NT	1.4
色氨酸	0.4	NT	0.2
缬氨酸	1.4	NT	NT

注：NT，未检测。

附表二 饲料原料的氨基酸组成

饲料原料的氨基酸组成（以干物质计，%）

项目	干物质重	可消化蛋白质	精氨酸	组氨酸	异亮氨酸	亮氨酸	赖氨酸	蛋氨酸	胱氨酸	苯丙氨酸	酪氨酸	苏氨酸	色氨酸	缬氨酸
脱水苜蓿草粉（17%粗蛋白）	92	17.10	0.77	0.33	0.81	1.28	0.85	0.27	0.29	0.80	0.54	0.71	0.34	0.88
螺旋藻粉	93	57.50	4.15	1.09	3.21	4.95	3.03	1.15	0.66	2.78	2.58	2.97	0.93	3.51
全大麦粉（带壳）	92	11.50	0.50	0.23	0.42	0.80	0.53	0.18	0.25	0.62	—	0.36	0.17	0.62
瞬间干燥猪血粉	92	87.60	3.70	7.00	1.51	12.30	8.22	1.67	0.82	5.48	2.80	3.18	1.00	7.70

（续）

项目	干物质重	可消化蛋白质	精氨酸	组氨酸	异亮氨酸	亮氨酸	赖氨酸	蛋氨酸	胱氨酸	苯丙氨酸	酪氨酸	苏氨酸	色氨酸	缬氨酸
喷雾干燥血粉	93	88.60	2.35	5.00	0.80	10.30	7.10	1.00	1.40	5.10	2.30	3.80	1.00	5.20
骨粉（鲑加工副产品）	92	39.50	2.56	0.66	1.13	2.04	2.06	0.91	0.27	1.12	0.81	1.36	0.29	1.50
脱水酒糟	92	23.10	1.27	0.52	1.54	2.49	0.09	0.46	0.35	1.44	1.20	0.93	0.37	1.61
亚麻籽粕	90	33.90	2.62	0.75	1.20	2.13	1.54	0.61	0.66	1.40	—	1.30	0.42	1.61
菜籽粕（溶剂提取）	93	38.00	2.32	1.10	1.51	2.60	2.02	0.77	0.97	1.50	0.99	1.50	0.46	1.94
双低菜籽粕	93	38.00	2.32	0.64	1.09	2.89	0.65	0.50	0.46	1.39	0.99	0.98	0.10	1.50
菜籽粕（低芥酸）	89	32.90	2.06	0.99	1.35	2.50	1.98	0.71	0.30	1.41	0.79	1.56	0.43	1.79
菜籽粕（低芥子油苷）	89	35.20	0.49	0.24	0.49	0.70	0.48	0.15	—	0.39	—	—	—	—
菜籽浓缩蛋白	90	69.20	4.20	1.70	2.82	4.92	3.10	1.26	1.27	2.81	0.99	2.49	0.20	3.33
酪蛋白（干燥）	91	84.00	3.26	2.82	4.66	8.79	7.35	2.70	0.41	4.79	4.77	3.98	1.14	6.10
椰子粕（机械提取）	93	22.00	2.30	0.30	1.00	1.49	0.54	0.33	0.20	0.80	—	0.60	0.20	1.00
玉米酒糟及可溶物（干燥）	91	27.00	1.10	0.65	1.00	2.80	0.90	0.51	0.50	1.20	0.31	0.92	0.20	1.33
玉米酒糟可溶物（干燥）	90	27.60	1.00	0.60	1.20	2.10	0.20	0.60	0.60	1.50	0.99	0.98	0.20	1.50
玉米蛋白粉（干燥）	90	21.50	1.04	0.67	0.66	1.96	0.63	0.35	0.46	0.76	0.58	0.74	0.07	1.01
玉米蛋白粉（干燥，粗蛋白＞60%）	91	63.70	1.90	1.20	2.30	9.40	1.07	1.90	1.10	3.80	0.87	2.00	0.30	2.70
玉米粉	88	10.20	0.40	0.25	0.29	1.00	0.26	0.18	0.19	0.42	—	0.30	0.07	0.42
普通黄玉米（粒状）	88	8.50	0.40	0.25	0.29	1.00	0.24	0.18	0.18	0.42	—	0.29	0.07	0.42
棉籽粕（溶剂萃取，41%粗蛋白）	92	41.70	4.18	1.07	1.45	2.32	1.60	0.58	0.73	2.18	0.94	1.34	0.53	1.90
蟹粉（干燥加工废弃物）	92	32.00	3.97	0.83	1.15	1.80	1.89	0.50	0.45	2.10	0.80	1.02	0.42	1.68
浓缩鱼膏（30%粗蛋白）	50	31.50	1.66	1.09	0.70	1.54	1.38	0.53	0.24	0.70	0.40	1.00	0.29	1.00
浓缩鱼膏（60%粗蛋白）	93	64.10	3.05	2.10	2.05	2.97	3.51	1.18	0.62	1.53	0.85	1.35	0.59	2.10
鳀粉（机械提取）	92	65.40	3.68	1.56	3.06	5.00	5.11	1.95	0.61	2.66	2.15	2.82	0.76	3.51
鲱粉（机械提取）	92	72.00	3.73	1.53	3.64	4.69	7.30	2.20	1.60	2.68	2.10	2.49	0.67	3.26

(续)

项目	干物质重	可消化蛋白质	精氨酸	组氨酸	异亮氨酸	亮氨酸	赖氨酸	蛋氨酸	胱氨酸	苯丙氨酸	酪氨酸	苏氨酸	色氨酸	缬氨酸
油鲱粉（机械提取）	92	64.50	3.66	1.78	2.57	4.54	4.81	1.77	0.57	2.51	2.04	2.64	0.66	3.03
鲑鱼粉（机械提取）	90	70.30	6.84	2.46	4.10	7.20	7.38	3.04	—	4.32	3.69	4.19	—	5.07
金枪鱼粉（机械提取）	93	59.00	3.43	1.75	2.45	3.79	4.22	1.47	0.47	2.15	1.69	2.31	0.57	2.77
白鱼粉（机械提取）	92	62.00	4.02	1.34	2.72	4.36	4.53	1.68	0.75	2.28	1.83	2.57	0.67	3.02
明胶	88	85.90	6.62	0.76	1.38	2.91	3.55	0.73	0.13	1.79	0.52	1.76	0.05	2.09
干海带粉	91	8.90	0.10	—	—	0.09	0.04	0.10	—	—	—	0.03	—	—
磷虾粉（带壳）	92	58.80	3.47	1.23	2.82	4.41	4.17	1.76	—	2.53	—	2.53	0.53	2.82
亚麻籽粕（溶剂提取）	90	35.00	2.97	0.68	1.56	2.06	1.24	0.59	0.59	1.57	1.03	1.26	0.52	1.74
羽扇豆粉	92	30.40	3.38	0.77	1.38	2.43	1.54	0.27	0.51	1.22	1.35	1.20	0.26	1.29
肉骨粉	94	50.90	3.60	0.96	1.70	3.20	2.60	0.67	0.33	1.70	1.30	1.70	0.26	2.25
猪肉骨粉	96	53.00	3.60	1.06	1.77	3.66	3.07	0.83	0.41	1.89	1.48	1.95	0.41	2.60
肉粉	93	55.60	3.60	1.14	1.60	3.84	3.07	0.80	0.60	2.17	1.40	1.97	0.35	2.66
脱壳燕麦	90	15.50	0.90	0.25	0.50	1.00	0.18	0.20	0.26	0.65	—	0.50	0.18	0.65
燕麦	90	11.20	0.70	0.18	0.43	0.81	0.39	0.17	0.19	0.52	0.46	0.36	0.15	0.56
豌豆浓缩蛋白	90	55.00	4.83	1.30	2.15	3.75	3.75	0.41	0.62	2.56	1.58	1.85	0.49	2.35
花生粕（溶剂提取）	92	49.00	3.37	0.96	1.43	3.00	2.67	0.65	0.50	1.70	1.09	1.65	0.30	2.45
豌豆（去皮，膨化）	90	25.30	1.87	0.54	0.86	1.51	1.50	0.21	0.31	0.98	0.71	0.78	0.19	0.98
马铃薯浓缩蛋白	93	81.10	4.02	1.77	4.62	7.94	6.09	1.77	1.27	5.19	4.56	4.62	—	5.33
饲料级禽肉骨粉	89	55.91	4.32	1.05	2.30	4.27	3.32	1.29	0.92	1.66	1.21	2.14	—	3.65
饲料级低灰分禽肉骨粉	89	62.13	4.99	1.17	2.38	4.14	3.44	1.24	—	1.63	1.30	2.39	—	3.03
宠物食品级禽肉骨粉	91	57.70	5.02	0.99	2.19	4.06	3.22	1.23	—	1.49	1.06	2.32	—	2.99
水解干燥羽毛粉	93	83.30	5.80	0.70	4.15	6.94	1.81	0.50	3.84	4.12	2.00	3.85	0.55	4.55
米糠	91	13.00	1.00	0.34	0.44	0.92	0.57	0.26	0.27	0.56	0.40	0.48	0.14	0.68
米糠（带胚，溶剂提取）	91	15.70	0.85	0.29	0.51	1.01	0.54	0.21	0.20	0.56	0.54	0.45	0.21	0.65
米糠	90	13.60	0.82	0.28	0.43	0.82	0.58	0.23	0.22	0.49	0.44	0.44	0.13	0.75
大米（打磨破碎，酿酒米）	91	7.37	0.55	0.18	0.36	0.64	0.27	0.18	0.13	0.36	—	0.27	0.09	0.46
大米浓缩蛋白	92	68.90	5.58	1.38	5.51	2.75	2.41	1.77	1.45	3.52	3.32	2.54	0.81	4.34

（续）

项目	干物质重	可消化蛋白质	精氨酸	组氨酸	异亮氨酸	亮氨酸	赖氨酸	蛋氨酸	胱氨酸	苯丙氨酸	酪氨酸	苏氨酸	色氨酸	缬氨酸
红花籽粕（机械提取）	91	20.00	1.20	0.48	0.28	1.10	0.70	0.40	0.50	1.00	—	0.47	0.30	1.00
红花籽粕（溶剂提取）	90	22.00	1.90	0.50	0.27	1.20	0.70	0.33	0.35	1.00	—	0.50	0.26	1.00
芝麻粕（机械提取）	94	42.00	5.06	1.16	2.28	3.30	1.37	1.48	0.60	2.32	—	1.71	0.82	2.53
干燥虾粉	88	39.50	1.54	0.51	0.91	1.62	1.66	0.55	0.59	0.99	—	3.24	0.24	1.03
高粱（粒装）	89	9.90	0.40	0.27	0.40	1.30	0.27	0.10	0.20	0.45	0.30	0.27	0.09	0.53
大豆粕（溶剂提取，无壳，48%粗蛋白）	90	48.50	3.60	1.30	2.60	3.80	2.24	0.70	0.71	2.70	1.25	2.00	0.70	2.70
大豆粕（溶剂提取，44%粗蛋白）	89	44.00	3.23	1.17	1.99	3.42	2.83	0.61	0.70	2.18	1.69	1.73	0.60	2.40
大豆浓缩蛋白	92	63.63	4.64	1.58	2.94	4.92	3.93	0.81	0.89	3.28	2.30	2.47	0.84	3.06
大豆分离蛋白	95	80.70	4.15	1.08	3.21	4.95	3.02	1.15	0.66	2.78	2.58	2.97	0.93	3.51
膨化大豆（全脂）	90	35.20	2.60	0.96	1.61	2.75	2.22	0.53	0.55	1.83	1.32	1.41	0.48	1.68
鱿鱼肝脏粉	89	45.20	2.48	—	—	—	2.56	0.97	—	—	—	1.74	0.49	—
鱿鱼粉	95	76.50	5.78	2.05	3.98	6.92	5.99	2.69	—	3.26	—	3.42	0.86	3.95
葵花籽粕（溶剂提取）	90	32.30	2.93	0.92	1.44	2.31	1.20	0.82	0.66	1.66	1.03	1.33	0.44	1.74
葵花籽粕（溶剂提取，无壳）	93	46.50	3.50	1.00	2.10	2.60	1.70	1.50	0.70	1.23	0.76	1.48	0.38	2.30
麦麸	89	14.80	0.64	0.30	0.51	0.89	0.36	0.21	0.27	0.63	0.43	0.37	0.17	0.59
小麦粉	88	11.70	0.86	0.39	0.51	0.92	0.58	0.19	0.26	0.55	0.38	0.46	0.25	0.69
小麦面筋粉	89	80.70	3.80	2.00	3.70	6.30	4.90	1.60	—	4.50	—	1.60	1.05	4.00
小麦次粉	89	16.60	0.97	0.44	0.70	1.10	0.70	0.12	0.19	0.50	0.29	0.51	0.20	0.75
硬红冬麦（粒装）	88	14.80	0.60	0.17	0.69	1.00	0.40	0.25	0.30	0.78	—	0.69	0.18	0.69
软白麦（粒装）	86	10.80	0.40	0.20	0.43	0.60	0.12	0.14	0.20	0.49	—	0.28	0.12	0.48
干燥乳清蛋白（低乳糖）	93	16.70	0.60	0.27	0.96	1.54	1.40	0.41	0.43	0.55	0.46	0.95	0.27	0.87
干燥啤酒酵母	93	42.60	2.20	1.09	2.15	3.13	3.22	0.74	0.50	1.83	1.55	2.20	0.56	2.39
干燥圆酵母	93	49.00	2.60	1.40	1.98	3.50	3.80	0.67	0.49	3.00	1.50	2.60	0.52	2.90

图书在版编目（CIP）数据

水产动物氨基酸营养研究/王连生，徐奇友主编．—北京：中国农业出版社，2022.8
ISBN 978-7-109-29777-7

Ⅰ.①水… Ⅱ.①王… ②徐… Ⅲ.①水产动物—氨基酸—动物营养—研究 Ⅳ.①S963

中国版本图书馆 CIP 数据核字（2022）第 140958 号

中国农业出版社出版
地址：北京市朝阳区麦子店街 18 号楼
邮编：100125
责任编辑：肖 邦　王金环
版式设计：杜 然　　责任校对：沙凯霖
印刷：北京大汉方圆数字文化传媒有限公司
版次：2022 年 8 月第 1 版
印次：2022 年 8 月北京第 1 次印刷
发行：新华书店北京发行所
开本：787mm×1092mm　1/16
印张：12.5
字数：312 千字
定价：65.00 元

版权所有·侵权必究
凡购买本社图书，如有印装质量问题，我社负责调换。
服务电话：010-59195115　010-59194918